**Design Center
Eval 5.4**

# PSpice for Windows
## A Primer

Edward Brumgnach, PE

**Delmar Publishers Inc.**™
I(T)P™ An International Thomson Publishing Company

New York • London • Bonn • Boston • Detroit • Madrid • Melbourne • Mexico City • Paris
Singapore • Tokyo • Toronto • Washington • Albany NY • Belmont CA • Cincinnati OH

**NOTICE TO THE READER**
Publisher does not warrant or guarantee any of the products described herein or perform any independent analysis in connection with any of the product information contained herein. Publisher does not assume and expressly disclaims any obligation to obtain and include information other than that provided to it by the manufacturer.

The reader is expressly warned to consider and adopt all safety precautions that might be indicated by the activities described herein and to avoid all potential hazards. By following the instructions contained herein, the reader willingly assumes all risks in connection with such instructions.

The publisher makes no representations or warranties of any kind, including but not limited to, the warranties of fitness for particular purpose or merchantability, nor are any such representations implied with respect to the material set forth herein, and the publisher takes no responsibility with respect to such material. The publisher shall not be liable for any special, consequential, or exemplary damages resulting, in whole or in part, from the reader's use of, or reliance upon, this material.

Cover design by Katie Hayden
**Delmar staff:**
Publisher: Mike McDermott
Administrative Editor: Wendy Welch
Project Editor: Barbara Riedell
Production Coordinator: Larry Main
Art & Design Coordinator: Lisa Bower
Assistant Editor: Jenna Daniels

For information, address
Delmar Publishers Inc.
3 Columbia Circle, Box 15015
Albany, NY 12203-5015

**COPYRIGHT © 1995 BY DELMAR PUBLISHERS INC.**
The trademark ITP is under license.

All rights reserved. No part of this work may be reproduced or used in any form, or by any means—graphic, electronic, or mechanical, including photocopying, recording, taping, or information storage and retrieval systems—without written permission of the publisher.

Printed in the United States of America
Published simultaneously in Canada
by Nelson Canada,
a division of The Thomson Corporation

1 2 3 4 5 6 7 8 9 10 XXX 00 99 98 97 96 95 94

**Library of Congress Cataloging-in-Publication Data**
Brumgnach, Edward.
  PSpice for Windows: A Primer / Edward Brumgnach.
    p.   cm.
  On CIP t.p.: "Design Center eval 5.4."
  Includes index.
  ISBN 0-8273-6821-6
  1. Design center for Windows. 2. Electric circuit analysis—Data processing.
3. Electronic circuit design—Data processing.
I. Title
  TK454.B765 1994
  621.319'2'02855369—dc20                                                94-3102
                                                                             CIP

*To Jean, Ian, Lara, and to my whole family
with gratitude and love!*

# Trademarks

"PSpice" is a registered trademark of MicroSim Corporation.

"The Design Center," "Probe," "Schematics," "StmEd," and "Stimulus Editor" are trademarks of MicroSim Corporation.

"Microsoft," "MS," and "MS-DOS" are registered trademarks, and "Windows" is a trademark of Microsoft Corporation.

"IBM" is a registered trademark of International Business Machines Corporation.

# CONTENTS

Foreword ix

## Chapter 1
**MICROSIM'S DESIGN CENTER FOR WINDOWS** 1
  1.1 System Requirements 1
  1.2 Software Installation 1
  1.3 Introduction 1
  1.4 The Schematics Editor 2
  1.5 The Schematics Editor Window 2
  1.6 The Mouse 4
  1.7 The First Exercise 5
  1.8 Continuing the Exercise 10
  1.9 Modifying the VSRC Symbol to Look Like a Battery 13

## Chapter 2
**OHM'S LAW AND POWER** 19
  2.1 Ohm's Law 19
  2.2 Power 19
  2.3 Given V and R, Find I 20
  2.4 Given Several Values of V, Keeping R Constant, Find Corresponding Values of I 22
  2.5 Given Several Values of R, Keeping V Constant, Find Corresponding Values of I 23
  2.6 Given I and R, Find V 25
  2.7 Given Several Values of I, Keeping R Constant, Find Corresponding Values of V 26
  2.8 Given Several Values of R, Keeping I Constant, Find Corresponding Values of V 27

## Chapter 3
**RESISTIVE CIRCUITS WITH DC SOURCES** 29
  3.1 Resistors in Series (DC Voltage Source) 29
  3.2 Resistors in Parallel (DC Current Source) 32

3.3  Resistors in Parallel (DC Voltage Source) 35
3.4  Resistors in Series-Parallel (DC Voltage Source) 36
3.5  Resistive Circuit with More than One DC Source 38
3.6  Loop Analysis 39
3.7  Superposition 41
3.8  Node Analysis 43
3.9  Thevenin's Equivalent Circuit and Norton's Equivalent Circuit 45

## Chapter 4
## IMPEDANCE CIRCUITS WITH AC SOURCES (PHASOR FORM) 51
4.1  AC Sources 51
4.2  Phasors and Complex Numbers 52
4.3  Impedance of a Resistor 53
4.4  Resistor with an AC Voltage Source 53
4.5  Impedance of a Capacitor 55
4.6  Capacitor with an AC Voltage Source 56
4.7  Impedance of an Inductor 57
4.8  Inductor with an AC Voltage Source 57
4.9  Impedances in Series with AC Voltage Source 58
4.10 Impedances in Parallel with AC Voltage Source 59
4.11 Impedances in Series-Parallel with AC Voltage Source 61

## Chapter 5
## RLC CIRCUITS WITH SINUSOIDAL SOURCES 65
5.1  Sinusoidal Voltage Source (VSIN) 65
5.2  Resistor with a Sinusoidal Voltage Source (Using PROBE) 65
5.3  Capacitor with a Sinusoidal Voltage Source (Using PROBE) 68
5.4  Inductor with a Sinusoidal Current Source (Using PROBE) 70
5.5  Series RC Circuit with a Sinusoidal Voltage Source (Using PROBE) 73
5.6  Series RL Circuit with a Sinusoidal Voltage Source (Using PROBE) 75
5.7  Series RLC Circuit with a Sinusoidal Voltage Source (Using PROBE) 79

Contents     vii

Chapter 6
**RLC CIRCUITS WITH PULSE VOLTAGE SOURCE (VPULSE)**    83
   6.1   Pulse Voltage Source (VPULSE)    83
   6.2   RC Circuit with VPULSE    83
   6.3   RLC Circuit with VPULSE    86

Chapter 7
**ELECTRONIC CIRCUITS**    87
   7.1   Half Wave Rectifier    87
   7.2   Full Wave Bridge Rectifier    88
   7.3   Full Wave Bridge Rectifier with Capacitor Filter    88
   7.4   Bipolar Transistor Amplifier Voltage Divider Bias    91
   7.5   An NPN BJT Amplifier    92
   7.6   Frequency Response of a JFET Amplifier    93
   7.7   Frequency Response of a Tone Control    95

Chapter 8
**DIGITAL CIRCUITS**    99
   8.1   The Digital Stimulus Source STIM1 and the TTL Inverter    99
   8.2   TTL 7408 AND Gate Logic    100
   8.3   Modulo 3 Synchronous Counter    102

Appendix A
**IMPEDANCE CIRCUITS WITH AC SOURCES (PHASOR FORM) (Chapter 4 for Version 5.3)**
   A.1   AC Sources    105
   A.2   Phasors and Complex Numbers    106
   A.3   Impedance of a Resistor    107
   A.4   Resistor with an AC Voltage Source    107
   A.5   Impedance of a Capacitor    110
   A.6   Capacitor with an AC Voltage Source (Using Include)    111
   A.7   Impedance of an Inductor    111
   A.8   Inductor with an AC Voltage Source (Using Include)    112
   A.9   Impedances in Series with AC Voltage Source (Using Include)    113
   A.10   Impedances in Parallel with AC Voltage Source (Using Include)    114
   A.11   Impedances in Series-Parallel with AC Voltage Source (Using Include)    116

## Appendix B
## MAKING AN OHMMETER (RMeter, RProbe)
## MAKING AN IMPEDANCE METER (ZMeter, ZProbe)     121

- B.1 Making an Ohmmeter (RMeter)     121
- B.2 Using RMeter     123
- B.3 Making RMeter into RProbe     123
- B.4 Making an Impedance Meter (ZMeter)     125
- B.5 Using ZMeter     126
- B.6 Making ZMeter into ZProbe     127

Index     131

# FOREWORD

MicroSim's Design Center for Windows is a computer environment which allows the simulation and analysis of analog and digital circuits. This computer environment is based on SPICE, which is an acronym for Simulation Program with Integrated Circuit Emphasis. SPICE was developed in the later 1960s at the University of California at Berkeley by Lawrence Nagel, with significant contributions by Ellis Cohen. SPICE was originally written in FORTRAN to simulate analog circuits on mainframe computers. With the proliferation and popularity of personal computers, several companies adapted the original program for desktop machines. MicroSim was one of these companies and the Design Center for Windows is the present culmination of the evolution of the PSpice program. MicroSim offers an evaluation version and the associated documentation at nominal prices (call MicroSim at 800-245-3022 for ordering information). The software is included with this book.

The documentation offered with the program is intended for practicing engineers and scientists and is not aimed at the beginner. The documentation consists of facts and "hints" which practicing professionals are then able to use to learn the inner workings of the program. Using the original documentation, the student—or for that matter any beginner—can learn to use this very powerful tool only after much perseverance and countless hours of using the program.

Design Center for Windows is a very powerful, somewhat complex, and often subtle tool. The purpose of this book is to guide the beginner through the mechanics of using the basic features of this program. The whole program is not explored and the author does not claim total knowledge of the product. Enough introductory and advanced features are examined, however, to put the reader well on his or her way toward working comfortably with PSpice for Windows. The book is meant to be used together and simultaneously with the Design Center for Windows program so that the features described in the book may be observed on the screen. Readers are encouraged to contact MicroSim and purchase the documentation for Design Center for Windows. The official documentation should be used as reference while going through this book and when the reader outgrows this book, the official documentation will serve as the user's guide to Design Center for Windows.

In order to examine features of Design Center for Windows, many electric circuits and electrical laws are used. This book is not a text in electric circuit analysis. Electric circuits are used and solved by traditional methods only to compare them with the Design Center for Windows solutions.

This book is:

- A good, *clear* introductory guide to the use of Design Center for Windows
- A review of electric circuit theory
- A review of electronics.

This book is not:

- A total guide to Design Center for Windows
- A text in electric circuits.

This book was written for anyone who wants to easily learn how to use PSpice for Windows. This book was meant to serve technical students, engineering students, physics students, or practicing technicians and engineers who want to learn PSpice for Windows without having to invest hours running the program while trying to learn its features. If one considers the documentation from MicroSim as "college," then the present book can be considered "grammar school" and "high school." The author expects the reader to graduate from this book to the official Design Center for Windows documentation from MicroSim.

The author would like to thank Mr. David Moretti for running the circuits on his computer to ensure that all the instructions were correct. A note of appreciation also goes to the MicroSim technical staff (especially Ray Hiveli) for their technical support.

# Chapter 1   MICROSIM'S DESIGN CENTER FOR WINDOWS

## 1.1   SYSTEM REQUIREMENTS

The Design Center Version 5.4 for Windows will run on any 80386- or 80486-based compatible PC with the following requirements: 640kB of low memory, a minimum of 4MB of extended memory, a floating point co-processor, a mouse, a serial port, either a 1.2MB 5.25" or 1.44MB 3.5" floppy, MS-DOS 3.0 or later, most displays supported by Microsoft Windows, and Microsoft Windows 3.1 or later.

Version 5.3 of the program does not need a math co-processor. If 5.3 is used, use Appendix A instead of Chapter 4. Microsoft Windows must be installed and running in either standard mode or 386 Enhanced Mode.

## 1.2   SOFTWARE INSTALLATION

Start Microsoft Windows. Insert the **Design Center** setup diskette labeled "Disk 1" into the floppy drive. Activate **Program Manager**. Choose **File/Run**. In the **Run** dialog box type $a:\backslash setup$ and press <enter> if you inserted the diskette in drive a (type $b:\backslash setup$ if you inserted the diskette in drive b). Follow the directions provided to complete the installation.

## 1.3   INTRODUCTION

MicroSim's Design Center for Windows is a computer environment which allows the simulation and analysis of analog and digital circuits. The Design Center for Windows consists of the following programs:

> *Schematics*: the graphical circuit editor (this is where you draw the circuit).
>
> *Probe:* the software oscilloscope.

1

*PSpice*: the circuit simulator.

*Stimulus Editor*: the stimulus generation tool.

*Parts*: the utility for creating semiconductor device model definitions.

*Filter Design*: an optional filter synthesis design aid.

This book emphasizes *Schematics*, *PSpice*, and *Probe*.

## 1.4 THE SCHEMATICS EDITOR

The Schematics Editor is the window that appears when *Schematics* is started. The Schematics Editor allows the user to construct circuit diagrams and also allows the circuit to be analyzed by *PSpice* and the resulting wave forms to be viewed through *Probe*. All actions are performed through selections from menus using the mouse and some keyboard keystrokes. The pull-down menus allow the placement of circuit parts, labels, text, and wires. Values of components may be changed and they may be moved or cut and pasted. Circuit diagrams may be copied to the Clipboard and pasted in other Windows applications.

## 1.5 THE SCHEMATICS EDITOR WINDOW (FIG 1-1)

The *title bar* (Fig 1-2) is displayed along the top of the window and contains the following information:

- The word *Schematics:*
- The file name of the circuit being worked on. A file name shown as <new> is assigned to a circuit diagram that either has not been assigned a file name or has not been saved yet
- p<#> next to the file name indicates the number of the page currently shown (the evaluation version allows only one page)
- part:<name> indicates the name of the part that was last retrieved from the library
- * next to the file name indicates that the file has not been saved since its last modification
- "(stale)" indicates that the circuit diagram was changed since its last simulation (Fig 1-2 and Fig 1-7)
- "(current)" indicates that simulation data is consistent with the present circuit diagram

CHAPTER 1    MicroSim's Design Center for Windows                              3

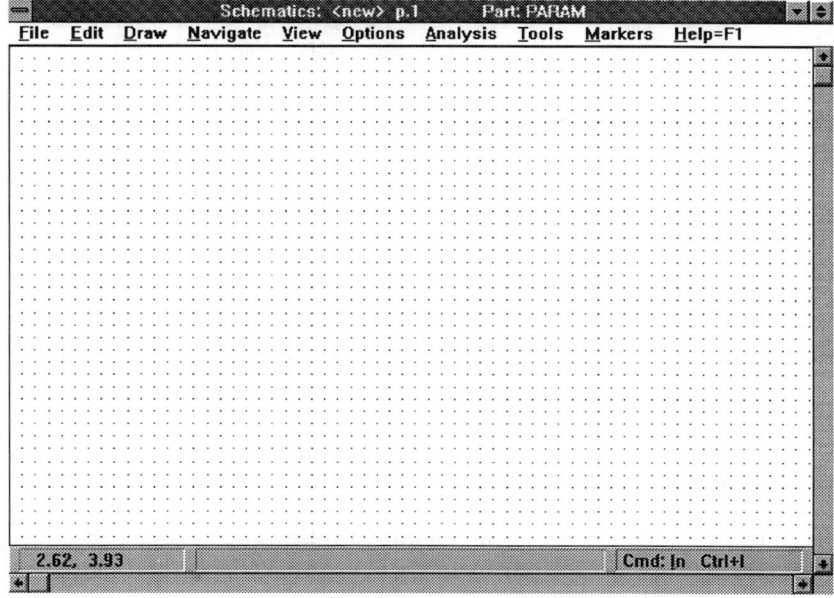

**1-1   Schematics' Editor Window**

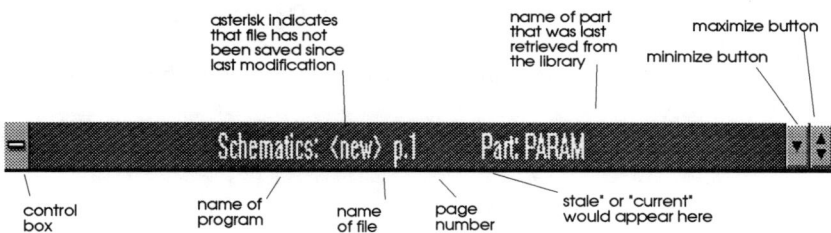

**1-2   Schematics' Title Bar**

The *menu bar* (Fig 1-3) appears immediately under the title bar. When items are selected from the menu, drop-down menus appear. Items may be selected from these drop down-menus by clicking the left mouse button on top of the item.

The *scroll bars* are located at the right and at the bottom of the window. Besides the usual manual scrolling, *Schematics* scrolls automatically when the cursor is moved outside the viewing area while moving or placing objects.

| File | Edit | Draw | Navigate | View | Options | Analysis | Tools | Markers | Help=F1 |

**1–3 Schematics' Menu Bar**

*Dialog boxes* appear when menu items followed by ellipses (...) are selected. These dialog boxes may contain all the standard elements: command buttons, option buttons, list boxes, text boxes, drop-down lists, and check boxes. After all the necessary information has been entered in a dialog box, click **OK** to carry out the command or click **Cancel** to cancel the command. Items in a dialog box that are grayed-out are not currently available for modification and cannot be selected.

The *status bar* (Fig 1–4) is located at the bottom of the window. The status bar may be activated or deactivated by selecting the proper check box in the options list box in the **Display Options** dialog box, accessed by choosing **Display Options** from the **Configure Menu**. The status bar provides the following information:

- The X and Y coordinates of the cursor (this feature may be activated or deactivated in the same dialog box as the status line)
- Prompts and warning messages
- The name of the command that will be executed if the repeat function is used

## 1.6 THE MOUSE

In *Schematics*, the mouse is used in conjunction with the keyboard to perform various functions. When the word "click" is used, the *left* mouse button is meant. If the right mouse button is to be clicked, it will

**1–4 Schematics' Status Bar**

CHAPTER 1    MicroSim's Design Center for Windows    5

be explicitly stated. A "click" consists of quickly pressing and releasing the mouse button. A "double click" consists of clicking twice quickly. "Pointing" consists of placing the mouse pointer directly on an object.

An object is "selected" by clicking on it. A selected object will turn red. Selecting a new object deselects the previous object. To select all objects within an area, select the area by holding down the left mouse button while "dragging" the mouse across the area. A "selection box" appears around the selected objects. Only objects entirely contained within the box are selected. An object is deselected by holding down the <Shift> key while clicking on the selected object. While an object is selected, an action may be performed on it (e.g., the object may be moved).

When the label or the value of a component is selected, a dotted "bounding box" appears around the object and around the attributes.

Several objects may be selected at the same time. Click on the first object, then hold down the <Shift> key and click on the additional objects to be selected.

To move an object or a group of objects, first select them by clicking. Then, while holding down the left mouse button, drag the objects to the desired location and release the mouse button.

To abort or cancel the current action, either press the <Esc> key, select **Cancel** if you are in a dialog box, or press the *right* mouse button.

## About Menu Selection and Clicking

Very often, clicking one menu choice opens another menu where another choice may be made, and so on. Instead of writing the word "click" several times, the options (separated by a slash) to be clicked in each subsequent menu will be listed. For example, the instruction: "click **Draw**, click **Get New Part**, click **Browse**" will be written as: "click **Draw/Get New Part/Browse**."

## 1.7  THE FIRST EXERCISE

The following exercise demonstrates some of the actions that are necessary to draw a circuit diagram and to work with the Design Center.

- Start *Windows*
- Start *Design Center Eval 5.4 (or 5.3)*
- Start *Schematics*

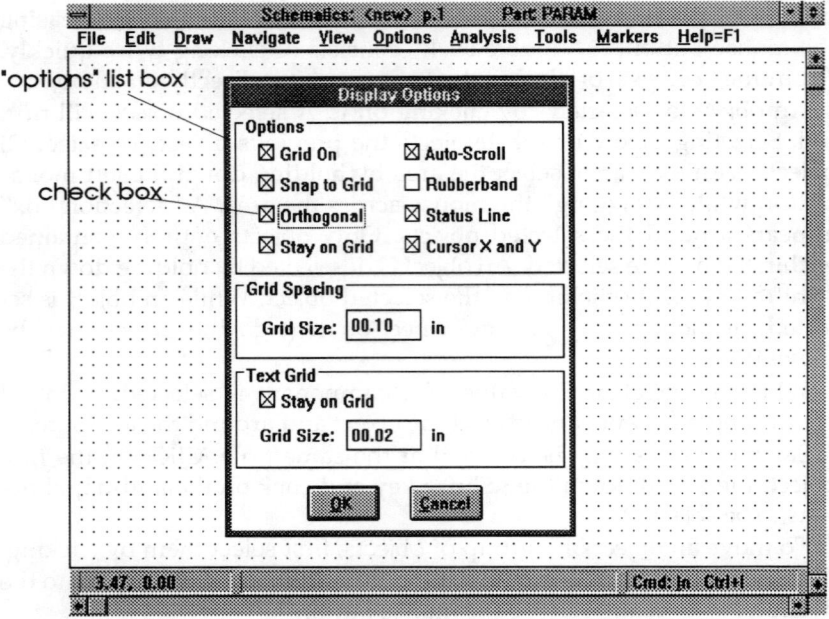

**1–5  The "Display Options" Dialog Box**

The Schematics Editor Window is now open. Some display options are now going to be activated.

- Click on **Configure/Display Options**

NOTE: An option is active when its check box has an X in it. An option is not active when its check box is empty. If an option is not active, it may be activated by moving the cursor on top of the option and clicking. If an option is active, it may be deactivated by clicking on the option. In other words, the option may be toggled on or off by clicking on it.

In the *Display Options* dialog box (Fig 1–5), find the *Options* list box and activate options by clicking in the appropriate check boxes: **Grid On, Snap to Grid, Orthogonal, Stay on Grid, Auto-Scroll, Rubberband, Status Line**, and **Cursor X and Y**. In the *Text Grid* list box, activate **Stay on Grid**. Click the **OK** command button. The drawing page in the Schematics Editor Window is now shown with a dotted grid and is now ready to receive the circuit diagram.

CHAPTER 1    MicroSim's Design Center for Windows    7

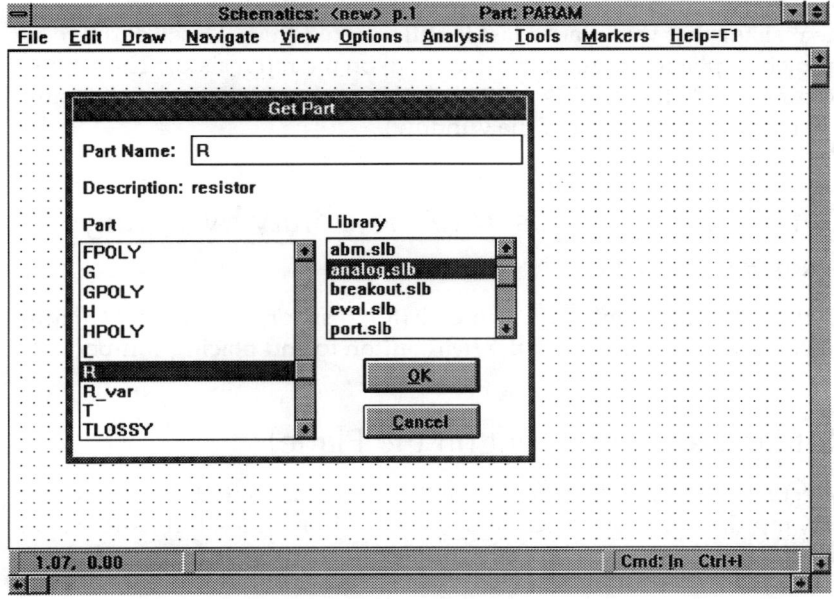

1–6  The "Get Part" Dialog Box

## Getting Components from the Library by Selecting from Menus

Click **Draw** to open the draw drop-down menu. In the Draw menu, click **Get New Part** to open the *Add Part* dialog box. In the *Add Part* dialog box, click the **Browse** command button to open the *Get Part* dialog box (Fig 1–6). In the *Get Part* dialog box, find the *Library* list box and click on the **analog.slb** option. Scroll the *Part* list box until the **R** (resistor) option appears. Click on the **R** option. Click on the **OK** command button. (Alternately, double click the **R** option). The cursor has now changed to an outline of a resistor.

(From now on, this will be written as: click **Draw/Get New Part/Browse/analog.slb/R/OK**.)

## Placing the Components

Click to place the resistor. Move the mouse and click again to place a second resistor. Click the *right* button to end placing resistors. Note that the resistor that was placed last is still selected (red). Deselect the resistor by first placing the cursor (arrow tip) on the red resistor. Then, while holding down the <Shift> key, click the mouse (alternately, click anywhere in an empty area of the window).

## Getting Components from the Library by Typing Their Acronyms

Click **Draw/Get New Part**. Type *R* (for resistor). Click **OK**. Click to place the component. Click the *right* button to end placing components.

## Rotating a Component (in the Plane)

Select R1. To rotate the component 90°, while holding down the <Ctrl> key, tap the <R> key. Alternately, select R1 and click **Edit/Rotate**. Deselect R1.

## Flipping a Component (Three-Dimensional Rotation)

Select R2. To flip the component, while holding down the <Ctrl> key, tap the <F> key. Alternately, select R2, and click **Edit/Flip**. Deselect R2.

## Selecting and Moving One Component

Click on R1. The resistor color changes to red. Place the cursor on top of the selected resistor and, while holding down the button, drag the resistor to another location and release the button. Deselect R1.

## Selecting and Moving More than One Component

Select R1. Move the cursor to R2. While holding down the <Shift> key, click on the resistor. Both resistors R1 and R2 are now selected. Both resistors may now be moved together. Place the cursor on top of either selected component and, while holding down the button, drag the resistors to another location and release the button. All selected components may be deselected simultaneously by clicking on an empty area. They may be deselected individually by holding down the <Shift> key while clicking on each one. Deselect both components.

## Changing the Component's Designation (Name)

Components are automatically designated by successive numbers (R1, R2, etc.) as they are placed. Their designation (name) may be changed. Double click on the designation R1. In the *text* box of the *Designation Reference* dialog box which is now open, delete the old name R1 and type the new name *Rx* and click **OK**. The new designation (name) of the resistor now appears as Rx. Change the designation of R2 to Ry. Change the designations back to R1 and R2.

## Changing the Component's Value

All resistors are placed with a value of 1k. This value may be changed. Double click on the value of R1. In the *text* box of the *Set Attribute* dialog box which is now open, type the new value (*2k*) and click **OK**. The new value of R1 now appears as 2k. Change the value of R2 to 3k.

## Moving the Position of the Component Name and/or Value

If the component name has to be moved, click on the name and release the button. Press the button again and, with the button depressed, drag the name box to the desired location. The value may be similarly moved. Move the names and values of R1 and R2 to the approximate locations shown in the circuit diagram (Fig 1–7).

## Placing Text on the Circuit Diagram

Click on **Draw/Text...** . In the *text* box of the *Place Text* dialog box which is now open, type the desired text (*2 W 5%*) and click **OK**. The outline of a box appears at the cursor. Move the cursor to the desired location (next to R1) and click. The text is placed next to resistor R1. Click the *right* button to end placing the text and click on an empty area to deselect the text box. Place the text *5W 5%* next to R2.

## Deleting Items

Move the cursor on top of R1 and look in the lower left-hand corner to see what its coordinates are. Move R1 to coordinates 4.4,2.6. Move R2 to coordinates 4.9,3. The bottom of resistor R1 should be at the same height as the top of R2 and the bottom of R2. Deselect all components. If necessary,

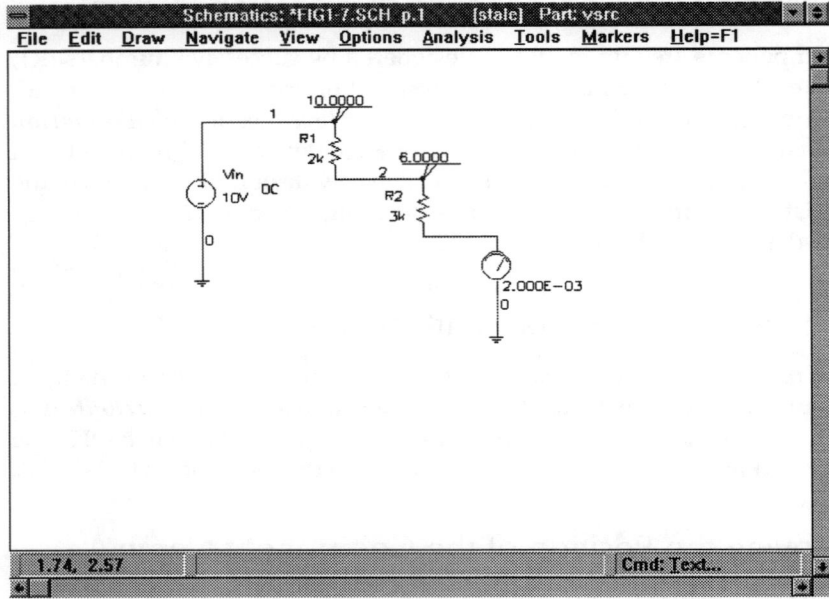

**1-7  Final circuit diagram showing voltages and currents**

scroll the window to place the circuit in the middle of the screen. This partial circuit diagram will be saved and completed imminently.

## Saving Files and Closing *Schematics*

Because this circuit diagram will be used later, it will now be saved. Click **File/Save**. Type *Fig1-7*. This will be the name of the file. Click **OK**. Close *Schematics* either by clicking **File/Exit** or by double clicking the Control Menu Box (upper left-hand corner). If the *Schematics* dialog box opens, click **YES** to save this particular diagram. *Schematics* is now closed.

## 1.8  CONTINUING THE EXERCISE
### Opening *Schematics* and Opening a File

Double click the **Design Center Eval** icon. Double click the *Schematics* icon. Click **Configure/Display Options** and activate all the options.

CHAPTER 1   MicroSim's Design Center for Windows                                    11

Click **File/Open**. Double click on **Fig1-7.sch**. The file is retrieved and the circuit diagram is shown on the monitor screen. You will now finish drawing the circuit diagram and then you will run PSpice to analyze it.

## Continuing Drawing the Circuit Diagram

- This circuit requires a 10V battery. Unfortunately, *Schematics* does not have a battery symbol (we will make a battery symbol later). A general purpose independent voltage source called VSRC is available. This source can be assigned a constant value. To access VSRC, click **Draw/Get New Part/Browse**. Scroll the *Library* list box until **source.slb** appears. Click **source.slb**. Scroll the *Part* list box until **VSRC** appears. Double click **VSRC**. Alternately, click **Draw/Get New Part**, type *VSRC*, and click **OK**. The source outline now appears as the cursor. Move it to approximately coordinates 3.5,3.1 and click to place the source. Click the *right* button to end placing this type of source. Note that the source designation is V1. To change the designation to Vin, double click on the designation, delete the designation V1, type the designation *Vin*, and click **OK**. To assign a 10V DC value to the source, double click on the source, double click **DC**, type *10*, click **Save Attribute/OK**. The source is now a 10 volt battery. By looking at the circuit diagram, however, you would never know this. It is a good idea to write this information next to the source. Click **Draw/Text...**, type *10V DC*, and click **OK**. Move the text box next to the source and click to place the text. Click the *right* button to end placement. Now it is obvious that this is a 10 volt DC source. Another way of showing the same information is to change the attributes of the symbol, as described in the next step.
- Click the text "**10V DC**" to select it. Tap <Del> to delete the text. Click the source to select it. Click **Edit/Symbol**. Click **Yes** if you are asked "Save changes to current page?". Click **Part/Attributes/DC/Display Value/Display Name/OK**. Click **Yes** to "Save changes to current attribute?". Select, drag, and drop **V?** to the upper right-hand side of the symbol. Select, drag, and drop **DC=** to the middle of the right side of the symbol. Click **Part/Save/File/Save/File/Return to Schematic**. Select the source and tap <Del>. Click **Draw/Get New Part**. Type *VSRC* and click **OK**. Place the source in the empty spot where you deleted the previous source. Click the right button. Double click the source designation, change it to "Vin", and click **OK**. Double click **DC=**, type

*10V*, and click **OK**. It should be obvious now that this is a 10V DC source. Later we'll learn how to make a symbol for a battery.
- We will want to measure the current in the circuit, so we have to insert an ammeter. *Schematics* has a DC ammeter called IPROBE. To access IPROBE, click **Draw/Get New Part/Browse**. Scroll the *Library* list box until **special.slb** appears. Click **special.slb**. In the *part* list box, double click **IPROBE**. Place IPROBE at approximately 5.4,3.4. Click the *right* button to end placement of IPROBE.
- To insert ground symbols (voltage reference points) in the circuit diagram, click **Draw/Get New Part/Browse/port.slb/GND_EARTH/OK**. Place one ground symbol under the source at approximately 3.5,3.8 and another ground under the ammeter at 5.4,4.0. Click the *right* button to end placing ground symbols.
- To interconnect the components with wires, click **Draw/Wire**. The cursor changes to a pencil shape. Click on top of the ground under the source, stretch the wire to the bottom of the source, and double click. Move the cursor to the top terminal of the source and double click the *right* button to resume wiring. Move the pencil-shaped cursor straight up to the level of the top terminal of R1 and click (this anchors the wire). Move the cursor to the top terminal of R1 and double click to suspend wiring. Move the cursor to the bottom of R1 and double click the *right* button to resume wiring. Move the cursor to the top of R2 and double click to suspend wiring. Similarly connect R2 to the ammeter and the ammeter to ground. When the wiring is completed, click the *right* button. Alternately, click to begin a new wire, stretch the wire to a desired location, click to anchor the wire, tap the space bar to suspend wiring, move the pencil to a new location, and click to resume wiring. Repeat the procedure until all the wiring is completed. Click the *right* button to end wiring.
- To facilitate discussion of a circuit, the circuit nodes should be numbered. In *Schematics*, wires are numbered instead of nodes. To assign a number to a wire, double click on the wire, type the wire number and click **OK**. Label as **1** the horizontal wire between the source and R1. Label as **2** the wire connecting R1 and R2. Label as **0** the wires on top of both ground symbols.
- *Schematics* provides a tool to measure the voltage from ground to any point in the circuit. This tool is called VIEWPOINT. We want to measure the DC voltage at nodes (wires) 1 and 2 from ground, so a VIEWPOINT has to be inserted at each one of these nodes. To insert VIEWPOINTS, click **Draw/Get New Part/Browse/spe-**

cial.slb/VIEWPOINT/OK. Place viewpoints at nodes 1 and 2. Click the *right* button to end placing viewpoints.
- To change the data in the name box of the page, scroll vertically and horizontally until the data box is on the screen. The upper left-hand corner of the box is at approximately 6,6. Double click in an empty area of the box and the *PartName* dialog box opens. Double click **Page Title=**, type *Fig 1–7* (which is the title you want), and click **Save Attr**. Similarly, insert *1* for PageNO, *1* for PageCount, and *1* for revision. For Company Name, I inserted "BRUMGNACH". You insert your name. In Line 1, I inserted "QUEENSBOROUGH COMMUNITY COLLEGE". In Line 2, I inserted our address "BAYSIDE, NY 11364". Then I scrolled down and in Line 3 inserted "718-631-6207" for the telephone number. In Date, I inserted 1/1/94. After each insertion you must click **Save Attr**. Click **OK**.
- To analyze the circuit, click **Analysis/Run PSpice**. The PSpice window will open and the analysis will be run. Once the analysis is completed, close the PSpice window. *Schematics* places the voltage values at each VIEWPOINT and the current value next to each ammeter (Fig 1-7).
- To obtain a hard copy of the circuit diagram and the results, click **File/Print/OK**.
- To save the circuit diagram, click **File/Save**.
- To close *Schematics*, either double click the *Control Menu Box* or click **File/Exit**.
- To close **Design Center Eval**, either double click the *Control Menu Box*, or click the *Control Menu Box* and click **Close**, or click **File/Close**.

## 1.9 MODIFYING THE VSRC SYMBOL TO LOOK LIKE A BATTERY (Skip If You Are Using Version 5.3)

In this section we will modify the symbol for the general purpose voltage source (**VSRC**) supplied by MicroSim to look like the more familiar BATTERY symbol.

First we must modify the msim.ini file to include a new symbol library. This new library will be named **Brum.slb**.

- Open **Design Center Eval/Schematics**.
- Click **Configure/Display Options/(turn on all options)/OK**.

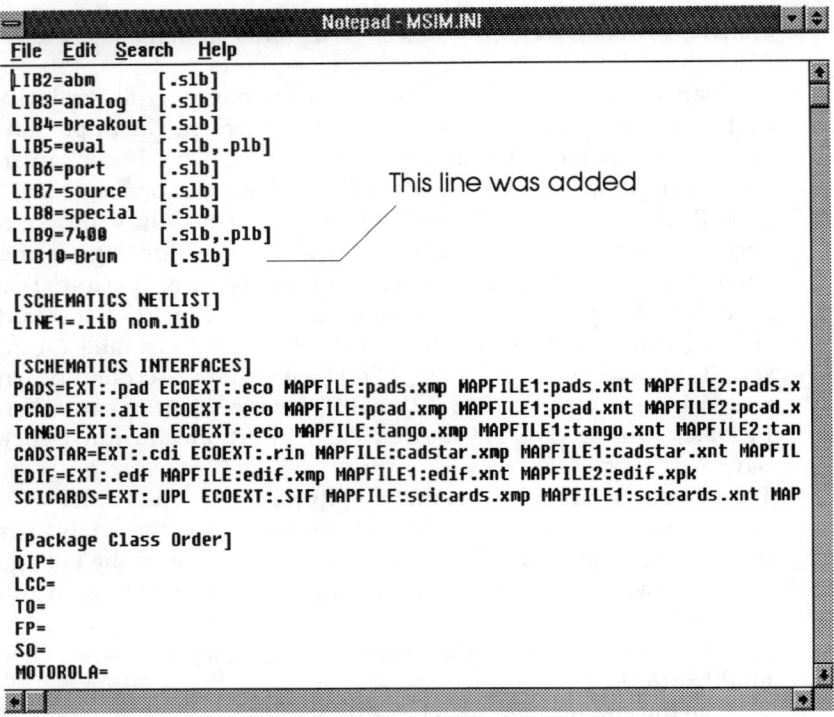

**1-8 Portion of the "win.ini" file with the Brum.lib added**

- In the following procedure we will modify the VSRC symbol to look like a battery symbol and we will add the Brum.slb library to the msim.ini file. Figure 1-8 shows a part of the msim.ini file with the new library added.

Next we will use the Library Editor to modify the symbol. Figure 1-9 shows the original symbol. Figure 1-10 shows the finished product.

- Click **File/Edit Library/Part/copy/Select Lib**. Scroll until **source.slb** is visible. Click **source.slb/OK**. Scroll the *Part:* box until **VSRC** is visible. Click **VSRC**.
- Click inside the *New Part Name:* box. Delete the contents of the box. Type *VDC*. Click **OK**.
- Click **Part/Attributes/refdes=V?** . In the *Value* box, click between the V and the question mark. Type *DC*. Click **Save Attr**.

CHAPTER 1    MicroSim's Design Center for Windows                              15

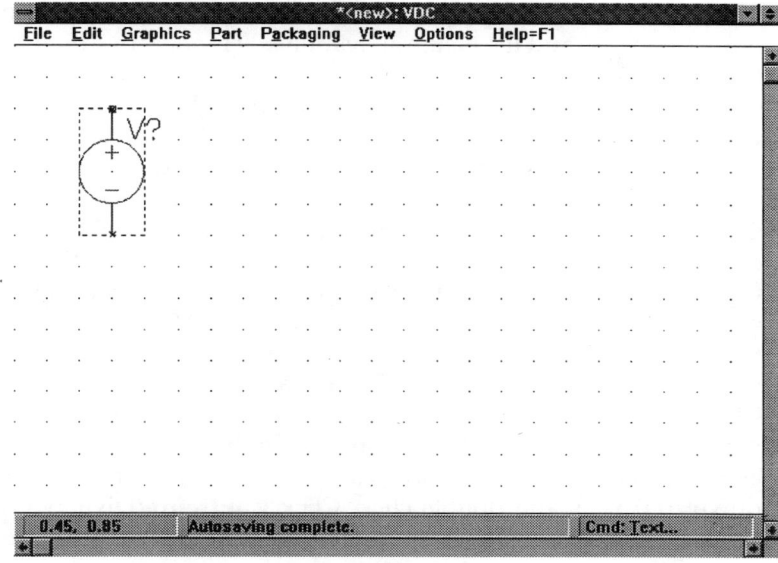

**1–9   Original source before editing**

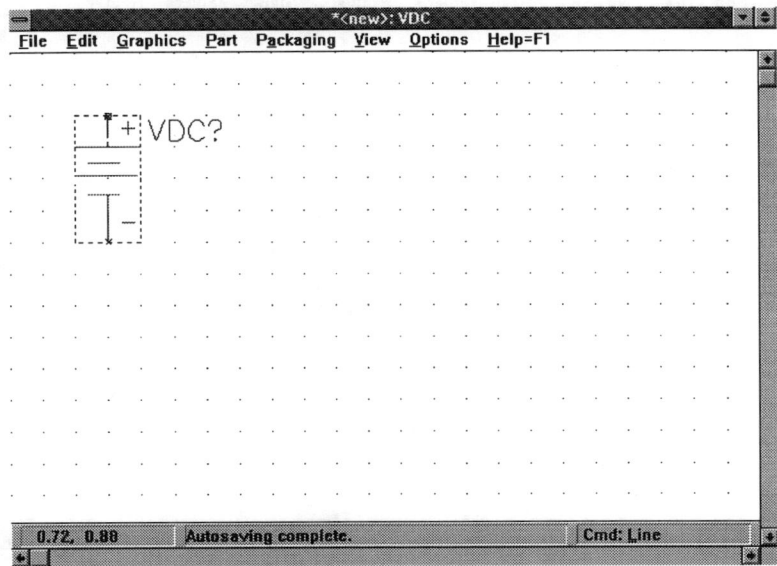

**1–10   Original source replaced by battery symbol**

- Click **DC=**. In the **Display Box**, make sure the **Value** activation box is active (X) and the **Name** activation box is not active (empty). Click **OK/YES**.
- Click, drag, and drop **VDC?** to the upper right of the symbol. Click, drag, and drop the plus sign above the circle to the right of the wire. Click, drag, and drop the minus sign below the circle to the right of the wire.
- Click on the circumference of the circle (the circle should change color). Tap <Del> to delete the circle. Click **Configure/Display Options** and deactivate (by clicking) **Snap to Grid** and **Stay on Grid**. Click **OK**. Click **Graphics/Line**. Click at 0.2,0.3, stretch the line over to 0.4,0.3, click and tap the <space bar>. Click at 0.25,0.35, stretch the line over to 0.35,0.35, click and tap the <space bar>. Click at 0.2,0.4, stretch the line over to 0.4,0.4, click and tap the <space bar>. Click at 0.25,0.45, stretch the line over to 0.35,0.45, click and tap the <space bar>. Click at 0.3,0.45, stretch the line down to 0.3,0.5, and double click. Click **Configure/Display Options** and reactivate (by clicking) **Snap to Grid** and **Stay on Grid**. Click **OK**.

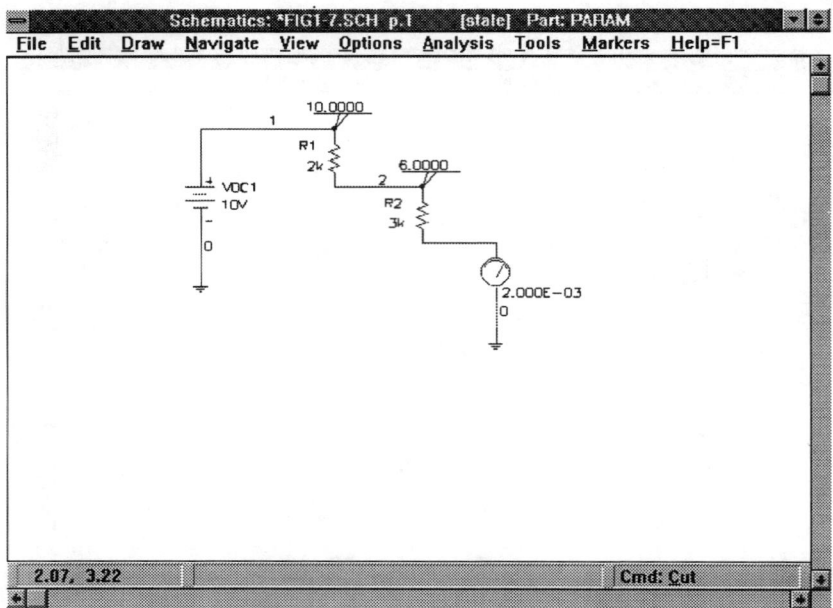

**1–11** Final circuit diagram with battery symbol and answers

CHAPTER 1    MicroSim's Design Center for Windows                          17

- Click **Part/Save Changes/File/Save As**. Type *Brum.slb*. Click **Yes/File/Return to schematics**. Exit *Schematics*.

Next we will use our newly created battery symbol.

- Open *Schematics*. Click **File/Open/Fig1-7.sch/OK**. Click the circle on the source symbol. Tap <Del> to delete the symbol.
- Click **Draw/Get New Part**. Scroll the **Library** box until **Brum.slb** appears. Click **Brum.slb/VDC/OK**.
- Place the battery symbol in the gap where the original source was removed. Double click the battery symbol. Double click **DC**. Type *10V*. Click **Save attr/OK**. Click, drag, and drop **10V** to the right of the battery symbol under **VDC1**.
- Click **Analysis/Run PSpice**. Close the **PSpice** window.
- The final circuit with the battery symbol and the answers is shown in Figure 1–11. The battery symbol may now be used in any circuit that requires a DC voltage source.
- Close *Schematics*. Click **Yes** to **Save current changes?**.

# Chapter 2    OHM'S LAW AND POWER

## 2.1  OHM'S LAW

In the mid 1800s, George Simon Ohm discovered the relationship between *voltage* (V), *current* (I), and *resistance* (R). This relationship is perhaps the single most important concept in circuit theory and is known today as Ohm's Law. Ohm's Law can be stated three ways.

If the voltage and the resistance are known, the current may be calculated from:

$$I = \frac{V}{R} \qquad \text{(Eq 2-1)}$$

If the current and the resistance are known, the voltage may be calculated from:

$$V = IR \qquad \text{(Eq 2-2)}$$

If the voltage and the current are known, the resistance may be calculated from:

$$R = \frac{V}{I} \qquad \text{(Eq 2-3)}$$

In all three cases:

>  V = voltage in volts (V)
>  R = resistance in ohms (Ω)
>  I = current in amperes (A)

## 2.2  POWER

*Power* is the rate at which energy is handled. In electrical systems, power is measured in *watts* and may be obtained by multiplying the voltage across an element by the current through the element.

$$P = V \cdot I \qquad \text{(Eq 2-4)}$$

If the power is related to a resistor, V=IR may be substituted into the equations, yielding:

$$P = I^2 \cdot R \quad \text{(Eq 2-5)}$$

Alternately, $I = \dfrac{V}{R}$ may be substituted into the equation, yielding:

$$P = \dfrac{V^2}{R} \quad \text{(Eq 2-6)}$$

In all three cases:

P = power in watts (W)
V = voltage in volts (V)
I = current in amperes (A)
R = resistance in ohms ($\Omega$)

## 2.3  GIVEN V AND R, FIND I

In modern electronics, DC energy sources are available as *constant DC voltage sources* (batteries) or *constant DC current sources*. A constant voltage source maintains a constant voltage between its terminals and produces a current according to Equation 2-1. A constant currrent source maintains a constant current from one point to another and produces a voltage between the two points according to Equation 2-2. In both cases R represents the total resistance as "seen" by the source.

To use *Schematics* to draw and analyze the circuit diagram shown in Figure 2-1, proceed as follows. Open *Schematics*. Click **Draw/Get New**

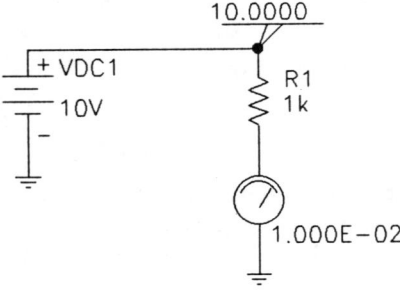

**2-1  Ohm's Law**

CHAPTER 2   Ohm's Law and Power                                             21

**Part/Browse/Analog/R/OK**. Rotate the resistor symbol to a vertical position, using <Ctrl> <R>. Place the resistor as shown in Figure 2-1. Click the right mouse button. Click **Draw/Get New Part/Browse/Brum/VDC/OK**. Place the battery symbol as shown in Figure 2-1. Click the right mouse button. Click **Draw/Get New Part/Browse/Special/Iprobe/OK**. Place the ammeter symbol right under and touching the resistor. Click **Draw/Get New Part/Browse/Port/GND_EARTH/OK**. Place a ground symbol touching the lower battery terminal and another ground symbol touching the lower ammeter terminal. Click the right mouse button. Click **Draw/Wire**. Click the pencil on the upper battery terminal, stretch the wire up a couple of dots, click; stretch the wire to the right so it lines up with the resistor, click; stretch the wire down to the upper resistor terminal and double click. Click **Draw/Get New Part/Special/Viewpoint/OK**. Place the viewpoint as shown in Figure 2-1. Click the right mouse button. Double click on the battery symbol. Double click on **DC**. Type *10V*. Click **Save Attribute/OK**. Click on **10V**. Click and drag the **10V** to just below the **VDC1** label, as shown in Figure 2-1. Double click on the resistor value (1k). Type *10k* and click **OK**. Click and drag the **10k** a little to the right. Click on **R1**. Click and drag **R1** above the **10k** as shown in Figure 2-1. Click on a clear area of the diagram. Click **Analysis/Run PSpice**. Type *Fig2-1* and click **OK**. The PSpice window appears and gives analysis information. When the analysis is completed, minimize the PSpice window. The circuit diagram is shown with the results. The viewpoint is showing 10 volts and the ammeter is showing 10 milliamps (1.000E-2 is PSpice's way of saying 1.000·10$^{-2}$ which is equal to 10mA).

This comfirms Ohm's Law as shown in Equation 2-1. The voltage is given as 10V. The resistance is given as 1kΩ. By dividing the voltage by the resistance, the current is obtained as 10mA. The energy source is supplying energy to the circuit at a rate of :

$$PS = V \cdot I = (10V) \cdot (10mA) = 100mW$$

The electrical energy is being converted to heat by the resistor at the same rate:

$$PR = V \cdot I = (10V) \cdot (10mA) = 100mW$$

This circuit diagram will be used in the next example.

## 2.4 GIVEN SEVERAL VALUES OF V, KEEPING R CONSTANT, FIND CORRESPONDING VALUES OF I

What would be the corresponding values of current if the voltage source varied in 1 volt steps from 0 volts to 10 volts while the resistance was held constant at 1kΩ. Obviously Equation 2-1 would have to be applied 11 times, and the resulting current values would be $I = \frac{0V}{1k\Omega} = 0, I = \frac{1V}{1k\Omega} = 1mA, I = \frac{2V}{1k\Omega} = 2mA$, and so on. The resulting current values may then be graphed against the values of the voltage.

*Schematics* may be used to solve this problem. To vary the source voltage from 0 to 10 volts in 1 volt steps, proceed as follows. Click **Analysis/Setup/DC Sweep/Voltage Source/Linear**. Click inside the **Name** box and type *VDC1*. Click inside the **Start Value** box and type *0*. Click inside the **End Value** box and type *10*. Click inside the **Increment** box and type *1*. Click **OK**. Click in the **Enable** square next to the **DC Sweep** button. Click **OK**.

To analyze the circuit, click **Analysis/Run PSpice**. When the analysis is completed, the PROBE window opens. To view the graph of the current versus the voltage, proceed as follows. Click **Trace/Add/I(V1)/OK**. The graph is shown in Figure 2-2. The annotations were placed on the

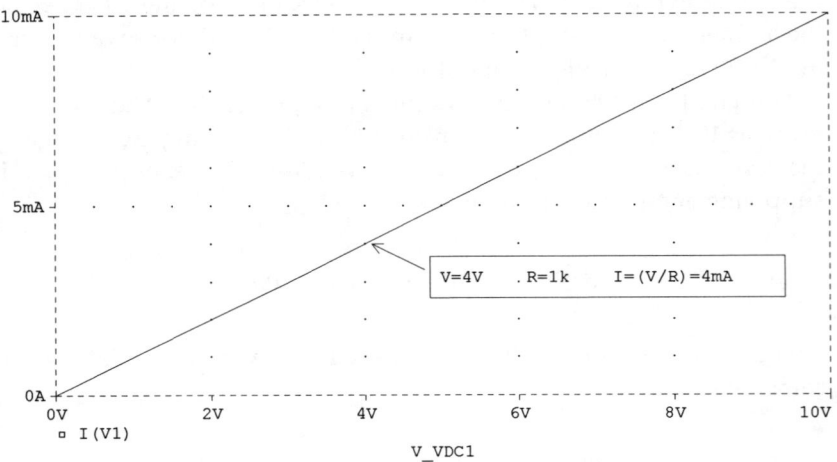

**2-2 Current variation as a result of voltage variation for constant resistance**

CHAPTER 2   Ohm's Law and Power                                     23

graph as follows. Click **Tools/Label/Text**. Type *V=4V R=1kΩ I=4mA* and click **OK**. Move text to the position shown in Figure 2–2 and click. To draw a box around the text, click **Tools/Label/Box**. Click where you want the upper left-hand corner of the box; move the mouse to where you want the lower right-hand corner of the box and click. To put an arrow from the text box to the point on the graph, click **Tools/Label/Arrow**. Click next to the text box to begin the arrow and click next to the desired point on the trace to end the arrow. To make the individual current values more apparent, place dots at the data points of interest. To do this, click **Tools/Options/Mark Data Points/OK**. A hard copy may be obtained as follows. Click **File/Print/OK**. The graph may be copied to the Clipboard (so that it may be used in a Windows word processor file by pasting) as follows. Click **Tools/Copy to clipboard**.

The graph clearly shows how the current varies from 0mA to 10mA as the voltage varies from 0V to 10V while the resistance is held constant at 1kΩ.

This circuit diagram will be used in the next example.

## 2.5   GIVEN SEVERAL VALUES OF R, KEEPING V CONSTANT, FIND CORRESPONDING VALUES OF I

What would be the corresponding values of current if the voltage source was kept constant at 10 volts but the resistor varied in 1kΩ steps from 1kΩ to 10kΩ? Obviously Equation 2–1 would have to be applied 10 times, and the resulting values of current would be: $I = \frac{10V}{1k\Omega} = 10mA$, $I = \frac{10V}{2k\Omega} = 5mA$, $I = \frac{10V}{3k\Omega} = 3.33mA$, and so on. The resulting current values may then be graphed against the values of resistance.

*Schematics* may be used to solve this problem. Refer to Figure 2–3. First the resistor value has to be made variable. Double click **1k**. Type *{RX}* and click **OK**.

Click **Draw/Get New Part/Browse/Special/PARAM/OK**. Place the **PARAMETER** symbol to the right of the resistor and click the right mouse button to end placement (see Fig 2-3). Double click **PARAMETERS**. Double click **Name 1** and type *RX*. Click **Save Attr**. Double click **VALUE 1** and type *1k*. Click **Save Attr/OK**. This procedure sets up the resistor value as a variable named RX which for now has a 1kΩ value. Next the resistor will be incremented in 1kΩ steps from 1kΩ to 10kΩ.

# CHAPTER 2  Ohm's Law and Power

```
         10.0000
   +VDC1    •      PARAMETERS:
   =10V    R1      RX      1k
    -     {RX}
   ⏚
          Ⓐ
         1.000E-02
          ⏚
```

## 2–3  Variable resistor (PARAMETER feature)

Click **Analysis/Setup/DC Sweep/Global Parameter/Linear**. Click inside the **Name** box and type *RX*. Click inside the **Start Value** box and type *1k*. Click inside the **End Value** box and type *10k*. Click inside the **Increment** box and type *1k*. Click **OK**. If the square **Enable** box next to **DC Sweep** does not have an X in it, click the box to enable the DC sweep. If the X is there, leave it alone. Click **OK**. The analysis is now set up. To run the analysis, click **Analysis/Run PSpice**. When the analysis is finished, the PROBE window opens up. Click **Trace/Add/I(V2)/OK**. The graph shows (see Fig 2–4) that as the resistance increases from 1k to 10k while the voltage is kept constant at 10 volts, the current decreases.

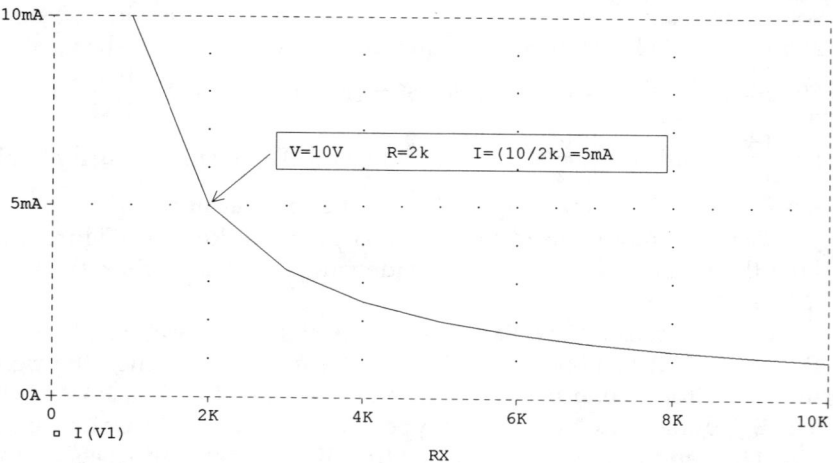

**2–4**  PROBE display of the current as a function of resistance

CHAPTER 2  Ohm's Law and Power  25

To show the data points, click **Tools/Options/Show Data Points/OK**. To get a hard copy of the graph, click **File/Print/OK**.

## 2.6 GIVEN I AND R, FIND V

A constant current source maintains a constant current between the nodes where the source is connected. The arrow inside the constant current source symbol shows conventional current direction. *Schematics* provides a constant current source named **ISRC**. Draw the circuit diagram shown in Figure 2–5.

Click **File/New**. Click **Yes** if you are asked whether you want to save the current diagram. Click **Draw/Get New Part/Browse/Analog.lib/R/OK**. Rotate the resistor with <Ctrl> <R>. Place the resistor as shown in Figure 2–5 and click the right mouse button to end placement of the resistor symbol. Click **Draw/Get New Part/Browse/Source/ISRC/OK**. Rotate the current source twice (<Ctrl> <R> <Ctrl> <R>) to make the current direction arrow point up. Place the current source as shown in Figure 2-5 and click the right mouse button to end placement of the source symbol. Double click the current source, double click **DC**, type *1mA*, and click **Save Attr/Change Display/Display Value/Display Name/OK/OK**. Click **DC=1mA**. Drag **DC=1mA** a little to the left. Click **I1**. Drag **I1** until it is below **DC=1mA**. Double click the resistor value **1k**, type *10k*, and click **OK**. Drag **10k** a little to the right. Click **R1**. Drag **R1** until it is above **10k**. Click **Draw/Get New Part/Browse/Port/GND_EARTH/OK**. Place one ground symbol touching the lower current source terminal and another ground symbol touching the lower resistor terminal. Click **Draw/Wire**. Click the pencil point on the upper source terminal, move the pencil up a couple of dots, and click. Move

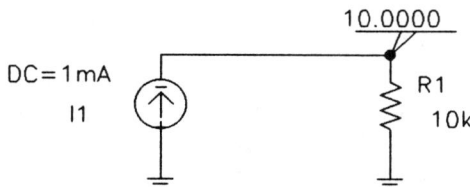

**2–5** A constant current source maintaining 1mA through a 10k resistor

the pencil to the right until it is in line with the resistor and click. Move the pencil down to the upper resistor terminal and double click. To put a voltage **VIEWPOINT** above the resistor, click **Draw/Get New Part/Browse/Special/VIEWPOINT/OK**. Place the **VIEWPOINT** as shown in Figure 2–5 and click the right mouse button to end placement. To analyze the circuit, click **Analysis/Run PSpice**. Type *Ohm2* and click **OK**. When the analysis is completed, minimize the PSpice window. As is apparent from Figure 2–5, the voltage across the resistor is 10 volts. This may be verified by applying Equation 2-2.

$$V = IR = (1mA)(10k\Omega) = 10V$$

## 2.7 GIVEN SEVERAL VALUES OF I, KEEPING R CONSTANT, FIND CORRESPONDING VALUES OF V

What would the corresponding values of voltage be if the current source was varied in 1mA steps from 0 to 10mA and the resistance was held constant at 1k? Obviously Equation 2-2 would have to be applied 11 times, and the resulting voltages would be: $V = IR = (0mA)(1k\Omega) = 0V$, $V = (1mA)(1k\Omega) = 1V$, $V = (2mA)(1k\Omega) = 2V$, and so on. The resulting voltage values may then be graphed against the current values.

*Schematics* may be used to solve this problem. To vary the value of the current source from 0 to 10mA in 1mA steps, proceed as follows. Click **Analysis/Setup/DC Sweep/Current Source/Linear**. Click inside the **Name** box and type *I1*. Click inside the **Start Value** box and type *0*. Click inside the **End Value** box and type *10m*. Click inside the **Increment** box and type *1m*. Click **OK**. Click the **Enable** square next to the **DC Sweep** button. Click **OK**. Double click the resistor value, type *1k*, and click **OK**. To label the wire connecting the current source and the resistor, double click the wire and type *1*. To analyze the circuit, click **Analysis/Run PSpice**. When the analysis is completed, the PROBE window opens. To view the graph of the voltage versus current, proceed as follows. Click **Trace/Add/V(1)/OK**. The graph with some annotations added is shown in Figure 2–6. The graph clearly shows that as the current varies from 0 to 10mA, the resulting voltage across a 1kΩ resistor varies from 0 to 10V.

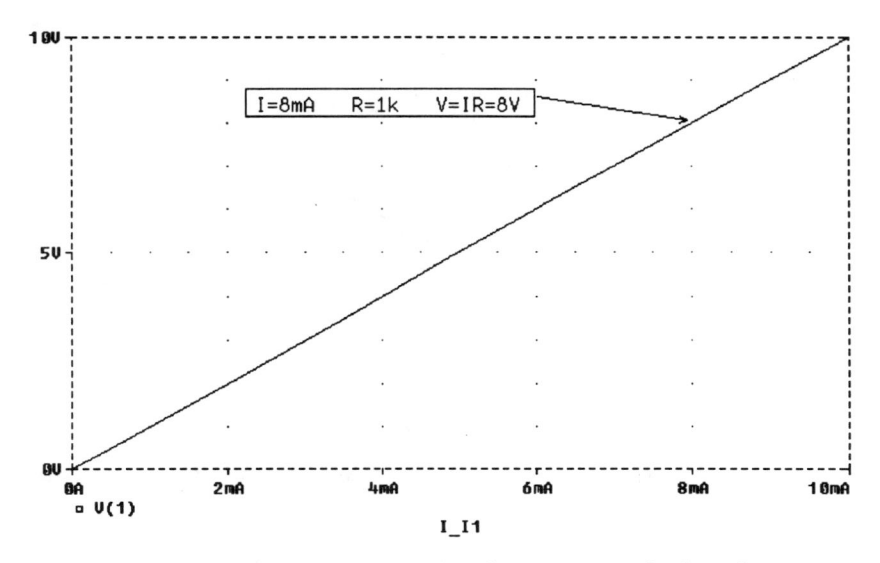

**2–6** Voltage variation as a result of current variation for constant resistance

## 2.8 GIVEN SEVERAL VALUES OF R, KEEPING I CONSTANT, FIND CORRESPONDING VALUES OF V

What would be the corresponding values of voltage if the current was kept constant at 1mA and the resistance varied in 1kΩ steps from 1 kΩ to 10kΩ? Obviously Equation 2-2 would have to be applied 10 times, and the resulting values of voltage would be: $V = IR = (1mA)(1k\Omega) = 1V$, $V = (1mA)(2k\Omega) = 2V$, and so on. The resulting voltage values may then be graphed against the values of resistance.

*Schematics* may be used to solve this problem. First the resistor value has to be made variable. Double click **1k**. Type *{RX}* and click **OK**.

Click **Draw/Get New Part/Browse/Special/PARAM/OK**. Place the **PARAMETER** symbol to the right of the resistor and click the right mouse button to end placement (see Fig 2-3). Double click **PARAMETERS**. Double click **Name 1** and type *RX*. Click **Save Attr**. Double click **VALUE 1** and type *1k*. Click **Save Attr/OK**. This procedure sets up the resistor value as a variable named RX which for now has a 1kΩ value. Next the resistor will be incremented in 1kΩ steps from 1kΩ to 10k. Click **Analysis/Setup/DC Sweep/Global Parameter/Linear**. Click in-

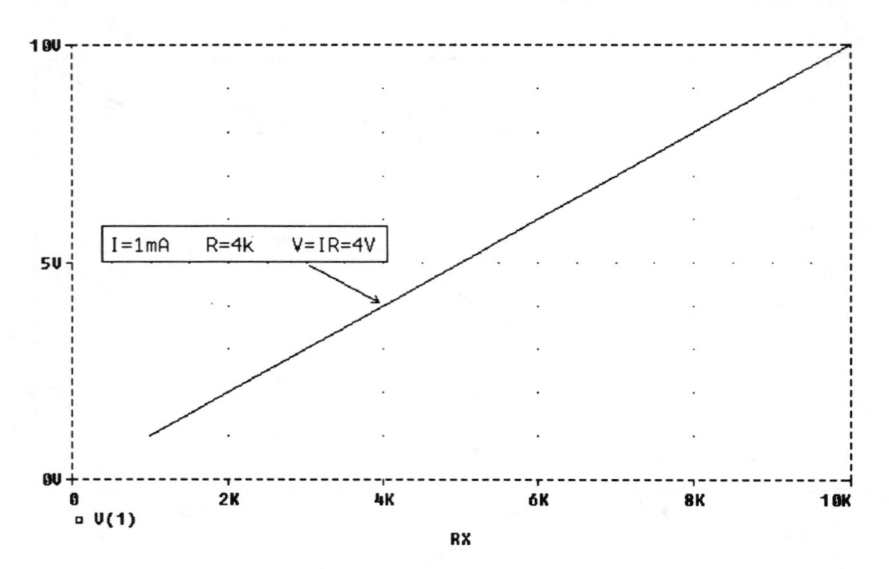

**2-7 PROBE display of the voltage as a function of increasing resistance with constant current**

side the **Name** box and type *RX*. Click inside the **Start Value** box and type *1k*. Click inside the **End Value** box and type *10k*. Click inside the **Increment** box and type *1k*. Click **OK**. If the square **Enable** box next to **DC Sweep** does not have an X in it, click the box to enable the DC sweep. If the X is there, leave it alone. Click **OK**. The analysis is now set up. To run the analysis, click **Analysis/Run PSpice**. When the analysis is finished, the PROBE window opens up. Click **Trace/Add/V(1)/OK**. The graph shows (see Fig 2-7) that as the resistance increases from 1kΩ to 10kΩ while the current is kept constant at 1mA, the voltage increases. To show the data points, click **Tools/Options/Show Data Points/OK**. To get a hard copy of the graph, click **File/Print/OK**.

Close *Schematics*, saving all circuit diagrams.

# Chapter 3 RESISTIVE CIRCUITS WITH DC SOURCES

## 3.1 RESISTORS IN SERIES (DC VOLTAGE SOURCE)

Resistors are in series if they are connected one after the other so that the *same* current goes through each one. Resistors in series have the following characteristics:

1. Resistors in series add
2. Resistors in series have the same current
3. Resistors in series divide the total voltage

Use *Schematics* to draw and analyze the circuit diagram shown in Figure 3–1. Figure 3–1 shows resistor R1 in series with resistor R2 (the three ammeters have no resistance). The total equivalent resistance is

$$RT = R1 + R2 = 2k\Omega + 3k\Omega = 5k\Omega$$

The constant voltage source VDC1 "sees" a total resistance of 5kΩ. Ohm's Law in the form of Equation 2–1 may be applied to find the total current coming out of the source:

$$I = \frac{VDC1}{RT} = \frac{10V}{5k\Omega} = 2mA$$

This is confirmed by the ammeters in Figure 3–1. The ammeters indicate 2.000E-03, which in PSpice talk means $2 \cdot 10^{-3}$. Because we are referring to a current, this is equivalent to 2mA. It should be apparent from the circuit diagram in Figure 3–1 that the current is the same through all the series components, since all three ammeters indicate the same current intensity. 2mA comes out of the positive terminal of the voltage source and it has no choice but to go through the first ammeter, where it is measured. It then goes through wire #1 and through R1. It then goes through wire #2 through another ammeter, where it is measured again.

**3-1 Series resistive circuit with DC voltage source**

It then goes through wire #3 and through R2. When it comes out of R2, the current goes through wire #4 and through another ammeter, where it is measured for the last time. The current then comes out of the ammeter and goes into ground. It then comes out of ground and goes into the negative terminal of the voltage source.

A current maintained through a resistor causes a voltage drop across the resistor. The voltage drop caused by a current through a resistor is always positive (+) at the terminal where the current enters the resistor and negative (−) at the terminal where the current leaves the resistor. The magnitude of the voltage drop may be obtained by Ohm's Law as stated in Equation 2-2 (V = IR). The voltage drops across R1 and R2 are, respectively:

$$VR1 = IR1 = (2mA) \cdot (2k\Omega) = 4V$$

$$VR2 = IR2 = (2mA) \cdot (3k\Omega) = 6V$$

Across each resistor, the voltage polarity is positive (+) on the top terminal and negative (−) on the negative terminal. The 6V drop across R2 is confirmed in Figure 3-1. The viewpoint at the bottom terminal of R2 indicates 0V. The viewpoint at the top terminal of R2 indicates +6V. This means that the top terminal of R2 is 6 volts higher than the bottom terminal. The voltage across R2 therefore must be 6 volts. The 4V drop

CHAPTER 3   Resistive Circuits with DC Sources                                    31

across R1 is also confirmed in Figure 3-1. The viewpoint at the top terminal of R1 indicates +10V. The viewpoint at the bottom terminal of R1 indicates +6V. This means that the top terminal of R1 is 4 volts higher than the bottom terminal. The voltage across R1 therefore is 4V.

At this point, it is easily seen that the 10V applied by the source across the two resistors got divided as 4V across R1 and 6V across R2. It should be obvious that the sum of the voltages across the two resistors is equal to the total applied voltage. This observation is originally credited to Kirchhoff and is known today as *Kirchhoff's Voltage Law* (referred to as KVL). The law says that **around any closed path, the sum of the voltage rises must be equal to the sum of the voltage drops**. A voltage rise is encountered when going through a voltage minus (–) to plus (+). A voltage drop is encountered when going through a voltage plus (+) to minus (–). Refer to Figure 3-1. Start at the ground at the negative terminal of the voltage source and proceed clockwise. Going up through the voltage source, we encounter a voltage rise of 10 volts because we go from the negative (–) to the positive (+) of the voltage. Going through R1, we drop from 10V to 6V (a 4V drop). Going through R2, we drop from 6V to 0V (a 6V drop). The 10V rise is equal to the sum of the 4V drop and the 6V drop. The ammeters are considered short circuits (zero resistance) and have no voltage drops across them.

$$10V = 4V + 6V$$

The voltage across each resistor may also be computed from a relationship known as the *voltage divider rule* (referred to as VDR):

$$VR1 = VT \frac{R1}{R1 + R2} \qquad \text{(Eq 3-1)}$$

In words, this says that the voltage across one of two series resistors is equal to the total applied voltage times the resistor of interest divided by the sum of the two resistors. For our example:

$$VR1 = (10V)\frac{2k\Omega}{2k\Omega + 3k\Omega} = (10V)\frac{2}{5} = 4V$$

$$VR2 = (10V)\frac{3k\Omega}{2k\Omega + 3k\Omega} = (10V)\frac{3}{5} = 6V$$

These two relationships show that the 2kΩ resistor always gets two-fifths of the applied voltage, while the 3kΩ always gets three-fifths of the applied voltage. The applied voltage may change, but the voltage in this voltage divider is always divided according to these ratios. The voltage divider circuit and concept are used often in electronics.

The voltage source power and the power to the resistors may be computed by applying Equation 2-4:

$$PS = V \cdot I = (10V) \cdot (2mA) = 20mW$$

$$PR1 = VR1 \cdot I = (4V) \cdot (2mA) = 8mW$$

$$PR2 = VR2 \cdot I = (6V) \cdot (2mA) = 12mW$$

It should be apparent that the rate at which the energy is converted to electrical form by the constant voltage source is the same rate at which the resistors are converting electrical energy to heat. This may be confirmed by observing that :

$$PS = PR1 + PR2$$

$$20mW = 8mw + 12mW$$

## 3.2 RESISTORS IN PARALLEL (DC CURRENT SOURCE)

Resistors are in parallel if they are connected between the same two nodes. A *node* is a junction where the terminals of different components may be connected. The usual way of showing two resistors in parallel is to use two parallel lines. R1//R2 denotes that R1 is in parallel with R2. Conductance (represented by upper case G) is measured in Siemens (represented by upper case S) and is the reciprocal of resistance.

$$G = \frac{1}{R} \qquad \text{(Eq 3-2)}$$

Resistors in parallel have the following characteristics:

1. Conductance values in parallel add
2. Resistors in parallel have the same voltage across them
3. Resistors in parallel divide the total current

Use *Schematics* to draw and analyze the circuit diagram shown in Figure 3-2. The current source is found in **source.slb** and is called **ISRC**. Once the source is placed, its value may be assigned by double clicking the source, double clicking **DC=**, typing *18mA*, and clicking **Save attr/OK**.

The conductance of each resistor may be computed by applying Equation 3-2:

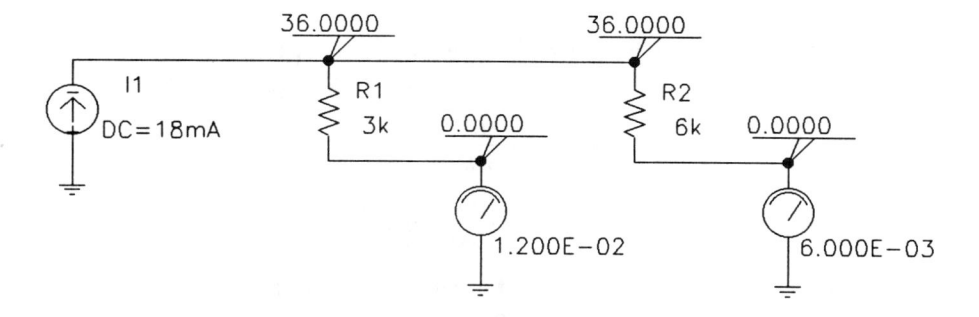

CHAPTER 3   Resistive Circuits with DC Sources                              33

```
         36.0000              36.0000
   I1      R1                   R2
  DC=18mA  3k    0.0000         6k    0.0000

                    1.200E-02           6.000E-03
```

## 3-2  Parallel resistive circuit with DC current source

$$G1 = \frac{1}{R1} = \frac{1}{6k\Omega} = 0.1667mS = 166.7\mu S$$

$$G2 = \frac{1}{R2} = \frac{1}{3k\Omega} = 0.3333mS = 333.3\mu S$$

The total conductance may be found by adding the individual conductances:

$$GT = G1 + G2 = 166.7\mu S + 333.3\mu S = 500\mu S$$

The total resistance is then the reciprocal:

$$RT = \frac{1}{GT} = \frac{1}{500\mu S} = 2k\Omega$$

Another way to find the equivalent resistance of two parallel resistors is to use the "product over sum" formula:

$$RT = \frac{R1 \cdot R2}{R1 + R2} \qquad \text{(Eq 3-3)}$$

Applying this formula, the preceding result may be confirmed:

$$RT = \frac{R1 \cdot R2}{R1 + R2} = \frac{(6k\Omega) \cdot (3k\Omega)}{6k\Omega + 3k\Omega} = \frac{18kk\Omega\Omega}{9k\Omega} = 2k\Omega$$

The 18mA current source therefore "sees" 2kΩ. The generated voltage may be found by applying Ohm's Law in the form of Equation 2-2:

$$V = IR = (18mA) \cdot (2k\Omega) = 36V$$

Figure 3-2 confirms this because the viewpoints at the top terminals of both resistors show 36V.

The current through each resistor may be found by Ohm's Law in the form of Equation 2-1:

$$IR1 = \frac{V}{R1} = \frac{36V}{3k\Omega} = 12mA$$

$$IR2 = \frac{V}{R2} = \frac{36V}{6k\Omega} = 6mA$$

At this point, it is easily seen that the total 18mA current divides into 12mA through R1 and 6mA through R2. It should be obvious that the sum of the two currents is equal to the total current to the two resistors. This observation is also credited to Kirchhoff and is known today as *Kirchhoff's Current Law* (referred to as KCL). The law says that **at any junction, the sum of the currents entering is equal to the sum of the currents leaving**. For our example:

$$8mA = 12mA + 6mA$$

The current through each resistor may also be found by applying the *current divider rule* (referred to as CDR):

$$IR1 = IT\frac{R2}{R1 + R2} \qquad \text{(Eq 3-4)}$$

In words, this says that the current through one of two parallel resistors is equal to the total current to the two resistors times the OPPOSITE resistor and divided by the sum of the two resistors. Applying this to our example:

$$IR1 = IT\frac{R2}{R1 + R2} = (18mA)\frac{6k\Omega}{6k\Omega + 3k\Omega} = (18mA)\frac{6k\Omega}{9k\Omega} = (18mA)\frac{2}{3} = 12mA$$

$$IR2 = IT\frac{R1}{R1 + R2} = (18mA)\frac{3k\Omega}{6k\Omega + 3k\Omega} = (18mA)\frac{3k\Omega}{9k\Omega} = (18mA)\frac{1}{3} = 6mA$$

These values are confirmed in Figure 3-2. The viewpoints at the top terminals of resistors R1 and R2 both indicate 36V, while the viewpoints at the bottom of both resistors indicate 0V. This means that the voltage across each resistor is 36 volts. In fact, it is the same 36V across both resistors. The ammeter below R1 indicated 1.200E-2, which in PSpice talk means $1.2 \cdot 10^{-2}$ which in turn is equivalent to $12 \cdot 10^{-3}$ which is 12mA because we are talking about a current. The ammeter below R2 indicates 6.000E-3, which means 6mA. These two indications confirm the expected currents through R1 and R2. These two values also confirm the CDR. These two relationships show that the 6kΩ resistor always gets one-third of the total current, while the 3kΩ always gets

CHAPTER 3  Resistive Circuits with DC Sources    35

two-thirds of the total current. The total current may change, but the current in this current divider is always divided according to these ratios. The current divider rule is often used in electronics.

The power for the current source and the resistors may be computed by Equation 2–4:

$$PS = V \cdot IT = (36V) \cdot (18mA) = 648mW$$

$$PR1 = V \cdot IR1 = (36V) \cdot (12mA) = 432W$$

$$PR2 = V \cdot IR2 = (36V) \cdot (6mA) = 216mW$$

It should be apparent that the source power is equal to the total resistor power (648mW = 432mW + 216mW). Comparing this to the results in section 3.1, it should be apparent that POWER ADDS, whether the components are in series or in parallel.

## 3.3  RESISTORS IN PARALLEL (DC VOLTAGE SOURCE)

Change the circuit diagram shown in Figure 3–2 by deleting the current source and adding a 36V DC voltage source and an ammeter to measure the total current. Run a PSpice analysis on the circuit. The resulting circuit diagram is shown in Figure 3–3.

The total resistance has not changed from the previous circuit, and an application of Ohm's Law shows that the total current is 18mA. This is confirmed in Figure 3–3 because the ammeter measuring the total current indicates 1.800E-2, which is 18mA. The rest of the circuit is as previously discussed.

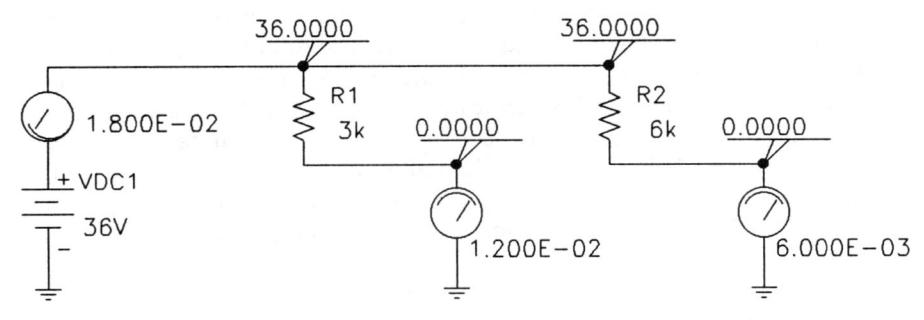

**3–3  Parallel resistive circuits with DC voltage source**

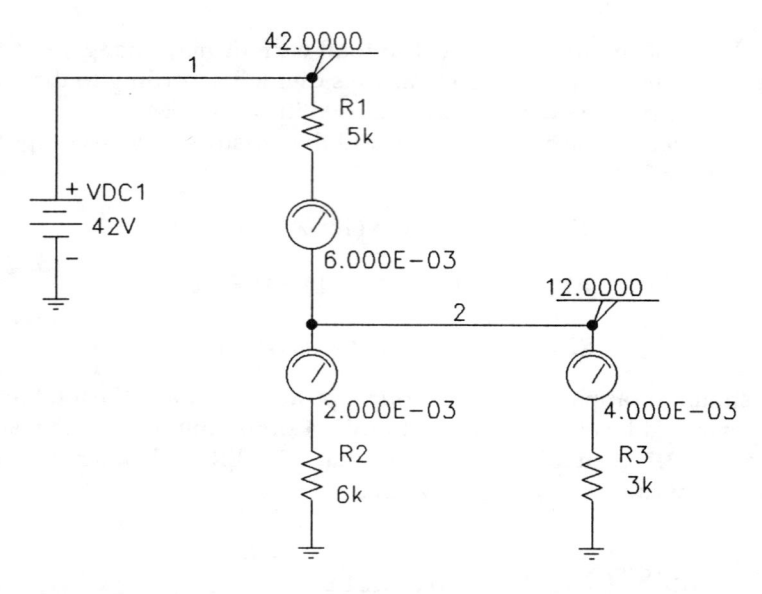

```
            42.0000
         1 ⟋
         ─────●
              ⋛ R1
              ⋚ 5k
   ╱┴╲         ⋛
  ─┬─ + VDC1  ⊘ 6.000E-03
   ┬   42V    │
  ─┴─ -       │         12.0000
   ╲┬╱        ●────────⟋──●
   ═           2         │
                         │
              ⊘ 2.000E-03  ⊘ 4.000E-03
              ⋛ R2         ⋛ R3
              ⋚ 6k         ⋚ 3k
```

**3–4 Series-parallel circuit with DC voltage source**

## 3.4 RESISTORS IN SERIES-PARALLEL (DC VOLTAGE SOURCE)

The circuit shown in Figure 3–4, although not a very practical circuit, is useful to demonstrate the concept of a series-parallel circuit. Resistor R1 cannot be combined with either resistor R2 or R3 because it is not in series or in parallel with either. Resistors R2 and R3 are in parallel because they are connected between the same two nodes. Recall from earlier in this chapter that 6kΩ//3kΩ is equal to 2kΩ. This parallel combination is then in series with R1. The total resistance "seen" by the voltage source is the sum of 5kΩ and 2kΩ. The total resistance is 7kΩ. This can be confirmed by using *Schematics* to draw Figure 3–5. Note that the resistance is obtained by dividing the voltage at the viewpoint by the value of the current source. Because the current source is equal to 1A, division by 1 does not alter the value of the numerator. The total resistance in ohms is the value at the viewpoint (7000 ohms or 7kΩ).

The 42 volt source puts out 6mA, according to Equation 2–1:

$$IT = \frac{VT}{RT} = \frac{42V}{7k\Omega} = 6mA$$

CHAPTER 3    Resistive Circuits with DC Sources                          37

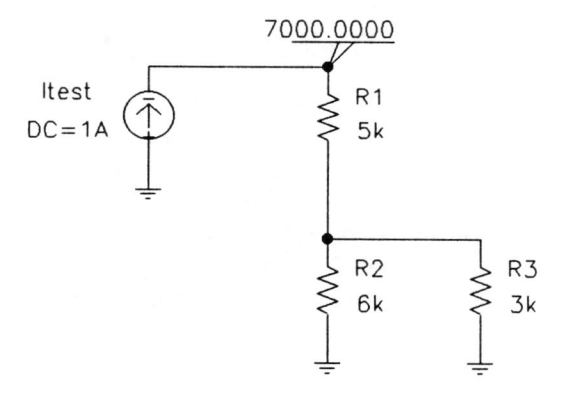

```
                    7000.0000
                       ╱
    Itest      ┬──────●────────┐
              ╱│╲              │
    DC=1A    ( ↑ )             ⌇ R1
              ╲│╱              ⌇ 5k
               │               │
               ⏚               │
                               ●────────┐
                               │        │
                               ⌇ R2     ⌇ R3
                               ⌇ 6k     ⌇ 3k
                               │        │
                               ⏚        ⏚
```

## 3–5 Total resistance obtained with *Schematics*

This current is confirmed in Figure 3–4 because the total current ammeter indicates 6.000E-3, which is 6mA. This current proceeds to wire #2, at which point it splits between R2 and R3. Using the CDR:

$$IR2 = IT \frac{R3}{R2 + R3} = (6mA) \frac{3k\Omega}{3k\Omega + 6k\Omega} = 2mA$$

The current through R3 may be obtained by applying Kirchhoff's Current Law (you could also apply the CDR again):

$$IR3 = IT - IR2 = 6mA - 2mA = 4mA$$

These currents are confirmed in Figure 3–4 because the ammeter in series with R2 indicates 2.000E-3 (2mA) and the ammeter in series with R3 indicates 4.000E-3 (4mA).

The voltage drops across each resistor may be found by applying Ohm's Law in the form of Equation 2–2:

$$VR1 = IT \cdot R1 = (6mA) \cdot (5k\Omega) = 30V$$

$$VR2 = IR2 \cdot R2 = (2mA) \cdot (6k\Omega) = 12V$$

$$VR3 = IR3 \cdot R3 = (4mA) \cdot (3k\Omega) = 12V$$

This shows that the voltage across the two parallel resistors is the same. Figure 3–4 confirms this because the viewpoint at wire #2 indicates 12V. The viewpoint at the top terminal of resistor R1 indicates 42V. The viewpoint at the bottom terminal of R1 indicates 12V. This implies that the top of R1 is 30 volts higher than the bottom terminal. In other words, the voltage across R1 is 30V.

## 3.5 RESISTIVE CIRCUIT WITH MORE THAN ONE DC SOURCE

Often we are asked to analyze a circuit with more than one source of energy. PSpice analyzes circuits with many energy sources as easily as it does circuits with one source. Predicting results by analyzing circuits with many sources without PSpice, however, can at times be pretty challenging. Depending on which results are desired, one of several methods of analysis is used. These methods of analysis are:

1. Loop analysis
2. Superposition
3. Node analysis
4. Thevenin's theorem
5. Norton's theorem

Loop analysis consists of applying KVL around each independent loop and then solving the resulting simultaneous equations for the loop currents. Superposition consists of finding the effects of each source one at a time and then superimposing those results. Node analysis consists of applying KCL at each node and then solving the resulting simultaneous equations for the node voltages. Thevenin's theorem consists of removing the load and finding an equivalent circuit made up of a voltage source in series with a resistance. The load is then put back and the load voltage and current may easily be found. Norton's theorem is similar to Thevenin's except that the equivalent circuit is made up of a current source in parallel with a resistance. The Thevenin resistance and the Norton resistance have the same value. When doing loop or node analysis by hand, it is very helpful and expedient to have a calculator that is capable of solving simultaneous equations. The Hewlett-Packard HP48 and the Texas Instruments TI85 are two machines that have this capacity. If these calculators are not used, some very tedious methods of solution of simultaneous equations have to be used. One of these methods is Cramer's rule, which involves the solution of several determinants.

Use *Schematics* to draw and analyze the circuit shown in Figure 3–6.

For this circuit, we are required to find the current through the load and the voltage across the load. PSpice for Windows (*Schematics*) analyzes the circuit and gives the results: 49.619V and 1.418A. We will use each of the five methods of analysis mentioned earlier to predict these results.

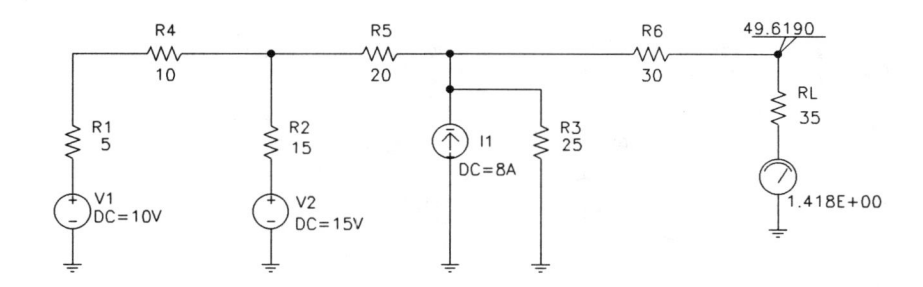

CHAPTER 3   Resistive Circuits with DC Sources                               39

**3–6 Resistive circuit with three energy sources**

## 3.6 LOOP ANALYSIS

When doing loop analysis, the task of obtaining the loop equations is somewhat simplified if all energy sources are voltage sources. A current source with its parallel resistor may be replaced by a voltage source with a series resistor. The value of the series resistor is the same as the parallel resistor. The value of the voltage source is the value of the current source multiplied by the resistor. In our example, I1 = 8A in parallel with R3 = 25Ω may be replaced with a voltage source of value VI1 = (8)(25) = 200V and a series R3 = 25Ω. The three loop currents—I1, I2, and I3—are drawn clockwise. According to theory, the currents may be drawn in any direction, but if they are drawn clockwise, the resulting equations achieve a very orderly format, as we will see later. The magnitude and polarity of the voltage or voltages (if more than one current) across each resistor are then specified in terms of that loop current. KVL is then applied around each loop, resulting in a set of simultaneous equations. These equations are then solved for the loop currents. Figure 3–7 shows the circuit with the current source replaced by the voltage source, the three loop currents drawn, and the magnitude and polarity of the voltage across each resistor specified.

In Figure 3–7, the voltage drop or drops across each resistor is labeled. R1 has a voltage drop from bottom to top of 5(I1), while resistors with two currents through them have two voltages across them. R2, for example (following the direction of I1), has a voltage drop of 15(I1) and a voltage rise of 15(I2). KVL is applied around each loop and the sum of the voltage rises and drops is found.

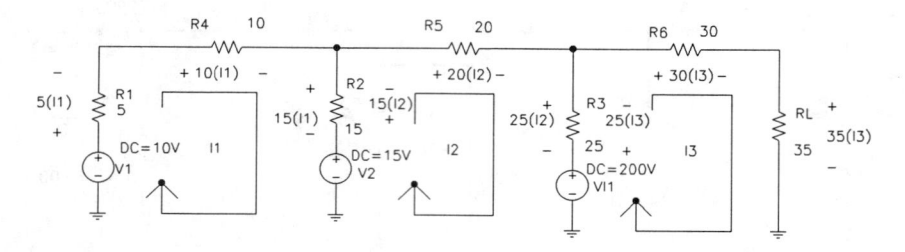

40                  CHAPTER 3    Resistive Circuits with DC Sources

**3–7 Circuit set up for loop analysis**

|       | Sum of voltage rises | Sum of voltage drops |
|-------|----------------------|----------------------|
| Loop 1 | 10 + 15(I2) | 5(I1) + 10(I1) + 15(I1) + 15 |
| Loop 2 | 15 + 15(I1) + 25(I3) | 15(I2) + 20(I2) + 25(I2) + 200 |
| Loop 3 | 200 + 25(I2) | 25(I3) + 30(I3) + 35(I3) |

Like terms are collected and the three equations are expressed in the following orderly format:

$$30(I1) - 15(I2) - 0(I3) = -5$$

$$-15(I1) + 60(I2) - 25(I3) = -185$$

$$0(I1) - 25(I2) + 90(I3) = 200$$

These three simultaneous equations may be easily solved with an appropriate calculator.

The results are I1 = –1.615A, I2 = –2.896A and I3 = 1.418A. A positive answer for a loop current indicates that the direction chosen for that loop current is correct; a negative answer indicates that the direction chosen for that current is incorrect. In this example, I1 and I2 have counterclockwise directions, while I3 is correctly shown clockwise. The desired current is I3. IL = I3 = 1.418A. The voltage across the load may be easily found by Ohm's Law. VL = (IL)(RL) = (1.418)(35) = 49.63V. These answers confirm the Windows PSpice solution.

In general, a three-loop circuit yields three simultaneous equations of the form

$$a_{11}I1 + a_{12}I2 + a_{13}I3 = b_1$$

$$a_{21}I1 + a_{22}I2 + a_{23}I3 = b_2$$

$$a_{31}I1 + a_{32}I2 + a_{33}I3 = b_3$$

If all the sources are constant voltage sources and all the loop currents

CHAPTER 3   Resistive Circuits with DC Sources                     41

have the same direction (clockwise, for example), then the coefficients may be obtained by observation as follows.

$a_{11}$, $a_{22}$, and $a_{33}$ are positive and each has a value equal to the sum of the resistors in loops 1, 2, and 3 respectively. All the other coefficients are negative and have a value equal to the resistance that is common to the two loops, denoted by the subscripts of the coefficient. $a_{12}$, for example, has the value of the resistance that is common to loops 1 and 2. If two loops do not touch (1 and 3 in this example), the coefficient with those two subscripts is zero. The constants (the $b$ terms) are obtained by following each current around the loop and ADDING voltage rises while SUBTRACTING voltage drops due to voltage sources.

For our example:

$a_{11} = 5 + 10 + 15 = 30$;  $a_{12} = -15$;  $a_{13} = 0$;  $b_1 = +10 - 15 = -5$
$a_{21} = -15$;  $a_{22} = 15 + 20 + 25 = 60$;  $a_{23} = -25$;  $b_2 = +15 - 200 = -185$
$a_{31} = 0$;  $a_{32} = -25$;  $a_{23} = 25 + 30 + 35 = 90$;  $b_3 = +200$

The three equations may then be written and solved:

$$30(I1) - 15(I2) - 0(I3) = -5$$

$$-15(I1) + 60(I2) - 25(I3) = -185$$

$$0(I1) - 25(I2) + 90(I3) = 200$$

## 3.7   SUPERPOSITION

In superposition, the circuit is analyzed with one source at a time. The effects of each source are then superimposed upon one another. In our example, there are three sources. We will use loop analysis to solve the three resulting circuits. The first circuit has V1 active. The other voltage sources are turned to zero (replaced by short circuits). The resulting equations are the same as before except for the constant terms.

$$30(I1) - 15(I2) - 0(I3) = 10$$

$$-15(I1) + 60(I2) - 25(I3) = 0$$

$$0(I1) - 25(I2) + 90(I3) = 0$$

Solving for I3, we get: I3 = 0.030A.

The second circuit has V2 active. V1 and V3 are turned to zero. The resulting equations are:

$$30(I1) - 15(I2) - 0(I3) = -15$$

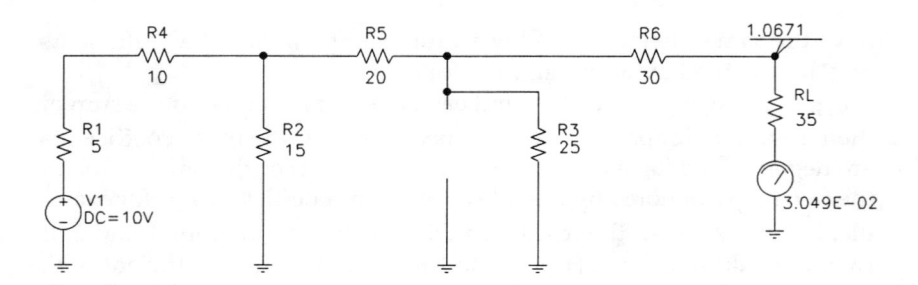

**3-8 Contribution of V1 to the load current**

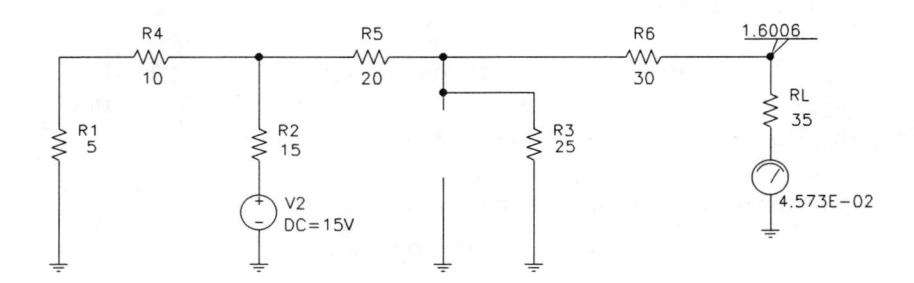

**3-9 Contribution of V2 to the load current**

$$-15(I1) + 60(I2) - 25(I3) = +15$$

$$0(I1) - 25(I2) + 90(I3) = 0$$

Solving for I3, we get: I3 = 0.046A.

The third circuit has V3 active. V1 and V2 are turned to zero. The resulting equations are:

$$30(I1) - 15(I2) - 0(I3) = 0$$

$$-15(I1) + 60(I2) - 25(I3) = -200$$

$$0(I1) - 25(I2) + 90(I3) = 200$$

Solving for I3, we get: I3 = 1.341A.

Superimposing the three values of I3, we get: 0.030 + 0.046 + 1.341 = 1.417A, which confirms the answer obtained by loop analysis.

*Schematics* may be used to solve the three individual circuits. The first circuit with V1 and the results is shown in Figure 3-8. The second circuit with V2 and the results is shown in Figure 3-9. The third circuit

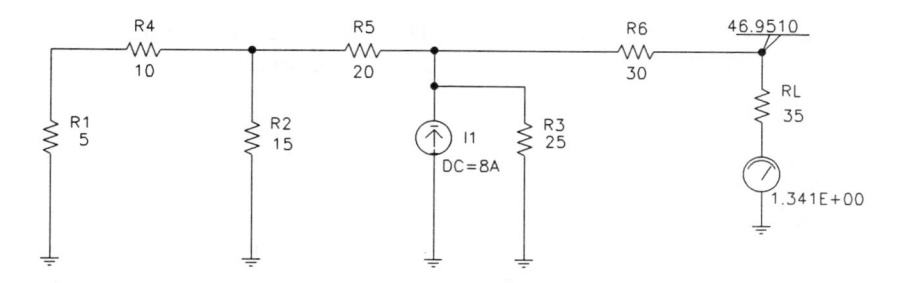

CHAPTER 3    Resistive Circuits with DC Sources                              43

**3–10  Contribution of the current source to the load current**

with V3 and the results is shown in Figure 3–10. It should be apparent that the three values of I3 compare favorably with the loop analysis solution (0.03049 + 0.0457 + 1.341 = 1.417A).

## 3.8  NODE ANALYSIS

Node analysis consists of obtaining a set of simultaneous equations by applying KCL to each node of the circuit. The equations are then solved to obtain the node voltages. The solution of circuit problems by node analysis is somewhat easier if all the sources of energy are in the form of constant current sources and all the resistances are expressed in terms of their conductance. It should be apparent in Figure 3–7 that R1 is in series with R4 and may be represented by RX = R1 + R4 = 5 + 10 = 15Ω. Also, R6 is in series with RL. This combination may be represented by RY = R6 + RL = 30 + 35 = 65Ω. Figure 3–11 shows our example with V1-RX replaced by IV1-RX and V2-R2 replaced by IV2-R2. In each case the voltage source in series with its resistance is converted to a current

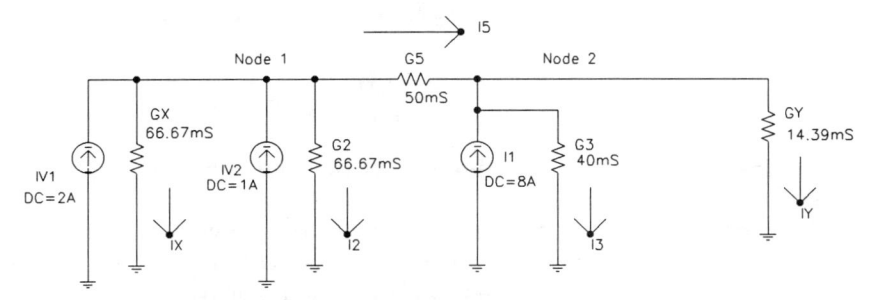

**3–11  Circuit set up for node analysis**

source in parallel with its resistance. The resistance value of the current source is the same as the resistance value of the original voltage source. The value of the current source is obtained by dividing the value of the original voltage source by the series resistance. All resistances are then converted to their conductance value. RX through RY are expressed in terms of their conductance GX through GY. Node 1 and node 2 are also labeled.

At each node, the entering currents are added together and the exiting currents are added together.

|        | Currents into node | Currents out of node |
|--------|--------------------|----------------------|
| Node 1 | 2 + 1              | IX + I2 + I5         |
| Node 2 | 8 + I5             | I3 + IY              |

Each current may be represented by the voltage across the conductor multiplied by the conductance:

$$IX = V1 \cdot GX = V1 \cdot (66.67mS)$$

$$I2 = V1 \cdot G2 = V1 \cdot (66.67mS)$$

$$I3 = V2 \cdot G3 = V2 \cdot (40mS)$$

$$IY = V2 \cdot GY = V2 \cdot (15.39mS)$$

$$I5 = (V1 - V2) \cdot G5 = V1 \cdot G5 - V2 \cdot G5 = V1 \cdot (50mS) - V2 \cdot (50mS)$$

KCL is applied. The currents entering each node are set equal to the currents leaving each node, yielding the following two simultaneous equations:

$$(183.34mS) \cdot V1 - (50mS) \cdot V2 = 1.667$$

$$-(50mS) \cdot V1 + (105.38mS) \cdot V2 = 8$$

Solving these yields the following answers: V1 = 34.224V, V2 = 92.154V. IL is the same current as IY, which may be found by Ohm's Law:

$$IL = IY = \frac{V2}{RY} = \frac{92.154V}{65\Omega} = 1.418A$$

This answer compares very well with the ones obtained by other methods.

In general, if all the sources of energy are converted to current sources and each resistance is represented by its conductance value, the node equations may be obtained by observation. The two equations are in the form:

# CHAPTER 3  Resistive Circuits with DC Sources

$$a_{11}V1 + a_{12}V2 = b_1$$

$$a_{21}V1 + a_{22}V2 = b_2$$

Because all sources are in current form and all resistances are expressed in conductance value, $a_{11}$ and $a_{22}$ are positive and their value is the sum of all the conductances connected to the node number designated by the subscript.

$$a_{11} = GX + G2 + G5 = 66.67mS + 66.67mS + 50mS = 183.34mS$$

$$a_{22} = G5 + G3 + GY = 50mS + 40mS + 15.38mS = 105.38mS$$

The values of $a_{12}$ and $a_{21}$ are negative and are the value of the conductance connecting nodes 1 and 2:

$$a_{12} = a_{21} = -G5 = -50mS$$

The value of the constant $b_1$ is obtained by ADDING all the current sources forcing current into node 1 and SUBTRACTING all current sources removing current from node 1. For our example:

$$b_1 = IV1 + IV2 = 0.667A + 1A = 1.667A$$

The value of the constant $b_2$ is obtained by ADDING all the current sources forcing current into node 2 and SUBTRACTING all current sources removing current from node 2. For our example:

$$b_2 = I1 = 8A$$

The two equations are (just as obtained previously):

$$(183.34mS) \cdot V1 - (50mS) \cdot V2 = 1.667$$

$$-(50mS) \cdot V1 + (105.38mS) \cdot V2 = 8$$

## 3.9 THEVENIN'S EQUIVALENT CIRCUIT AND NORTON'S EQUIVALENT CIRCUIT

Both Thevenin's equivalent circuit and Norton's equivalent circuit supply the load with the same current and voltage as the original circuit. Thevenin's equivalent circuit consists of a voltage source in series with a resistance, whereas Norton's equivalent circuit consists of a current source in parallel with the same resistance. The value of the Thevenin voltage source (VTH) is the value of the voltage that would be measured at the load terminals if the load was removed. The value of the Norton current (IN) is the value of the current through a short circuit

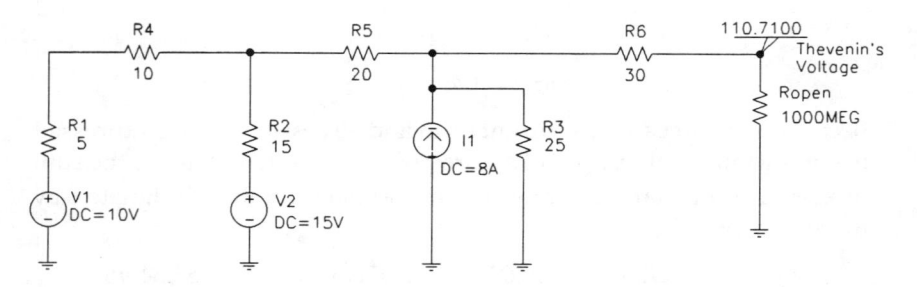

**3–12 Circuit set up for Thevenin's voltage**

that would be put at the load terminals after the load was removed. The value of both the Thevenin resistance and the Norton resistance is the value of the resistance that would be measured between the load terminals if the load was removed and all energy sources inside the circuit were turned off (voltage sources get replaced by short circuits and current sources get replaced by open circuits). It turns out that the equivalent resistance for both Thevenin's and Norton's circuits is the ratio of the Thevenin voltage and the Norton current. Since *Schematics* can only figure out voltages and currents, this last observation is the way to obtain the equivalent resistance. Figure 3–12 shows the circuit used to obtain Thevenin's voltage. The load is replaced by a 1000M resistor instead of an open circuit, because in PSpice open circuits cannot be used. The Thevenin voltage is 110.71V.

Figure 3–13 shows the circuit used to obtain the Norton current, which is 2.569A.

The equivalent resistance may be found as follows:

$$RTH = RN = \frac{VTH}{IN} = \frac{110.71V}{2.569A} = 43.1\Omega$$

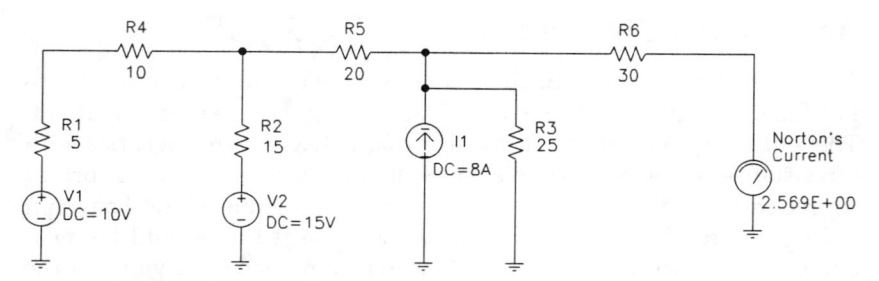

**3–13 Circuit set up for Norton's current**

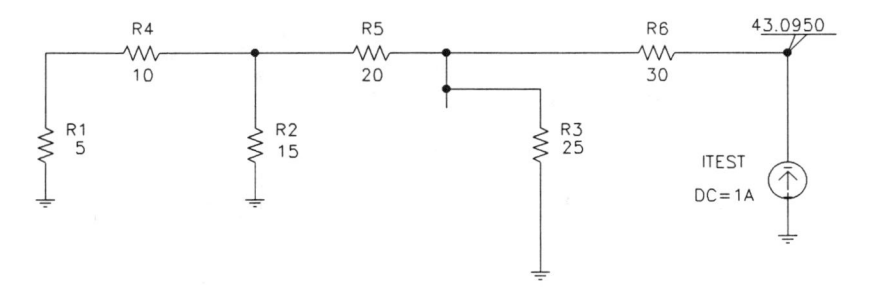

CHAPTER 3   Resistive Circuits with DC Sources                47

[Circuit diagram showing R4=10, R5=20, R6=30 in series across top; R1=5, R2=15, R3=25 as shunts to ground; ITEST DC=1A current source with reading 43.0950]

## 3-14 The viewpoint shows Thevenin's and Norton's resistance in ohms

Thevenin's and Norton's resistance may also be found by using *Schematics* and "Brumgnach's one amp rule." Figure 3-14 shows the circuit diagram drawn with constant voltage sources set to zero (replaced by short circuits) and constant current sources also set to zero (replaced by open circuits). The load is removed and replaced by a 1A constant current source. A viewpoint is placed on top of the current source. This viewpoint measures the voltage. Thevenin's and Norton's resistance may be obtained by applying Ohm's Law R = V/I. Because the current is 1A, the resistance (in Ω) is the same numerical value as the voltage (dividing by one leaves the numerator unchanged). The viewpoint then shows the Thevenin and Norton resistance in ohms.

Thevenin's equivalent may be drawn as shown in Figure 3-15. Norton's equivalent may be drawn as shown in Figure 3-16.

The load voltage and current obtained by these two methods compare very well with those obtained by other methods.

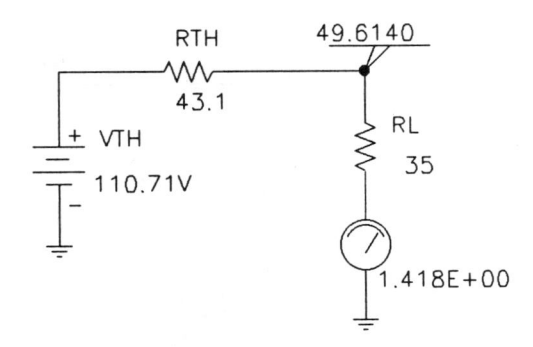

[Circuit diagram: VTH = 110.71V, RTH = 43.1, RL = 35, voltage reading 49.6140, current reading 1.418E+00]

## 3-15 Thevenin's equivalent circuit

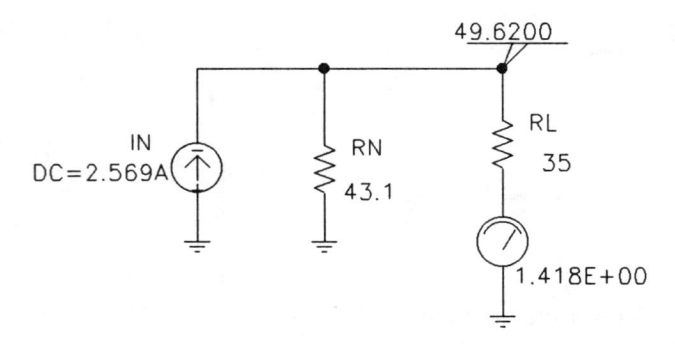

**48**  CHAPTER 3   Resistive Circuits with DC Sources

```
                          49.6200
                          
         IN          RN        RL
  DC=2.569A          43.1      35
                               
                               1.418E+00
```

**3-16  Norton's equivalent circuit**

The same equivalents may be obtained by hand. Figure 3-17 shows the circuit with the load removed and all energy sources shut off. The resistance from A to B may be found as follows. R1 is in series with R4 (5 + 10 = 15). This combination is in parallel with R2 (15//15 = 7.5). This combination is in series with R5 (7.5 + 20 = 27.5). This combination is in parallel with R3 (27.5//25 = 13.1). Finally this combination is in series with R6 (RTH = RN = 13.1 + 30 = 43.1). This is the result obtained previously.

Figure 3-18 shows the circuit used to obtain VTH. Since the load is open and there is no current through R6, there is also no voltage dropped across R6. The voltage on the right terminal of R6 is the same as the voltage on the left terminal of R6. If the current I2 is found, then the voltage across R3 may be obtained by Ohm's Law and VTH is equal to the combination of the 200V source and the voltage across R3. I2 may be found by loop analysis.

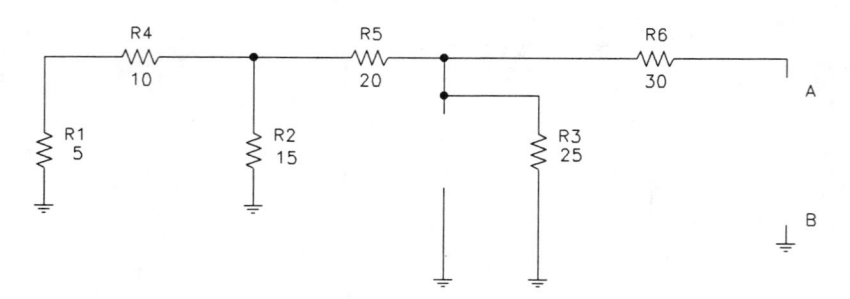

```
         R4           R5            R6
         10           20            30
                                         A
    R1          R2           R3
    5           15           25
                                         B
```

**3-17  Network used to find RTH and RN**

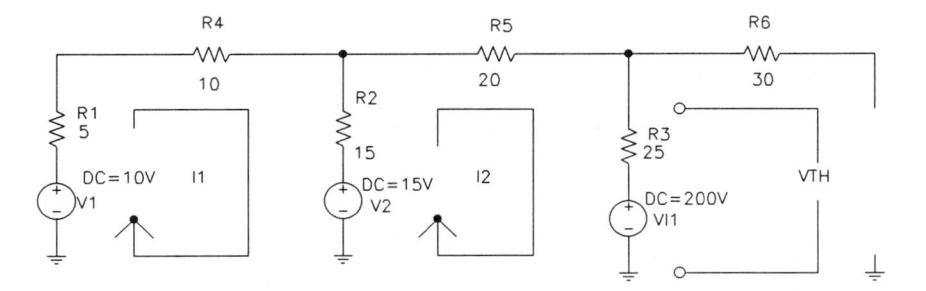

CHAPTER 3    Resistive Circuits with DC Sources                          49

**3-18   Circuit used to obtain VTH**

The two loop equations are:

$$30I_1 - 15I_2 = -5$$

$$-15I_1 + 60I_2 = -185$$

The current I2 is obtained via calculator as −3.571A. The minus sign indicates that the current is opposite the direction shown for I2. The current is therefore up and the voltage drop across R3 is developed with the positive at the bottom terminal of R3 and the negative at the top terminal of R3. The magnitude of the voltage drop is obtained by Ohm's Law and is

$$3.571A \cdot 25\Omega = 89.275V$$

The Thevenin voltage is then obtained by subtracting this voltage from the 200V source.

$$V_{TH} = 200V - 89.257V = 110.725V$$

This compares favorably with the answer obtained earlier.

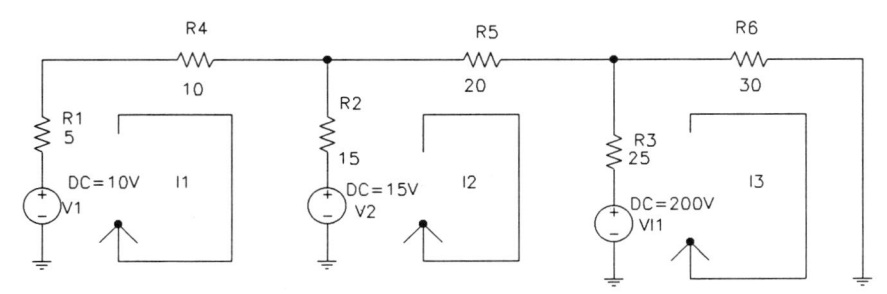

**3-19   Circuit used to obtain IN**

Figure 3–19 shows the circuit used to obtain the Norton current. The three loop equations are:

$$30I_1 - 15I_2 + 0I_3 = -5$$
$$-15I_1 + 60I_2 - 25I_3 = -185$$
$$0I_1 - 25I_2 + 55I_3 = 200$$

The current $I_3$, which is the Norton current, is obtained with an appropriate calculator as 2.569A, which is equal to the Norton current obtained earlier.

# Chapter 4 IMPEDANCE CIRCUITS WITH AC SOURCES (PHASOR FORM)

(If your computer does not have a math co-processor and you are using Version 5.3 of PSpice for Windows, use Appendix A instead of this chapter.)

## 4.1 AC SOURCES

Electrical energy is usually generated and distributed by means of alternating currents and voltages. The general form of an alternating current waveform is:

$$i = I_{peak}\sin(\omega t \pm \varphi)\ amps \qquad \text{(Eq 4-1)}$$

Similarly for an alternating voltage:

$$v = V_{peak}\sin(\omega t \pm \varphi)\ volts \qquad \text{(Eq 4-2)}$$

Where:

$i$ = instantaneous current (A)
$I_{peak}$ = peak or maximum current (A)
$\omega$ = radian frequency (rad/s)
$\varphi$ = phase angle (degrees)
$v$ = instantaneous voltage (V)
$V_{peak}$ = peak or maximum voltage (V)

Although these waveforms are sinusoidal, it is customary to refer to them by their *phasor notation*, which consists of their EFFECTIVE value and PHASE ANGLE. The effective (or RMS) value for a sinusoid is: RMS = 0.707 • *Peak*. The equation for the typical house line voltage is:

$$v = 170\sin(377t + 0°)\ volts$$

This voltage is usually referred to as $120\angle 0°V$. The 170V peak is 120V RMS and the phase angle is 0 because this waveform is considered reference. The frequency (60 Hz or 377 rad/s) is not included in the phasor notation because the circuits considered do not alter the frequency. The phasor form of an electrical quantity is represented by placing a bar over its abbreviation. For our example:

$$\overline{V} = 120\angle 0° \text{ volts}$$

## 4.2 PHASORS AND COMPLEX NUMBERS

Phasors are the polar form of complex numbers. Complex numbers may be represented in "polar form" (as a magnitude and an angle) or in "rectangular form" (as a real part and an imaginary part). The two forms are:

$$\overline{V} = c \angle \Theta \text{ volts} \quad \text{(Eq 4-3)}$$

$$\overline{V} = (a + jb) \text{ volts} \quad \text{(Eq 4-4)}$$

One form may be converted to the other by simple trigonometry and the Pythagorean theorem. If the polar form is known, then:

$$a = c \cdot \cos\Theta \qquad b = c \cdot \sin\Theta$$

If the rectangular form is known, then:

$$c = \sqrt{a^2 + b^2} \qquad \Theta = \tan^{-1}\left(\frac{b}{a}\right)$$

Complex number arithmetic applies to phasors. For addition and subtraction, it is simpler to use the rectangular form.

$$(a_1 + jb_1) + (a_2 + jb_2) = (a_1 + a_2) + j(b_1 + b_2)$$

$$(a_1 + jb_1) - (a_2 + jb_2) = (a_1 - a_2) + j(b_1 - b_2)$$

For multiplication and division, it is simpler to use the polar form.

$$(c_1 \angle \Theta_1) \cdot (c_2 \angle \Theta_2) = (c_1 \cdot c_2) \angle(\Theta_1 + \Theta_2)$$

$$\frac{(c_1 \angle \Theta_1)}{(c_2 \angle \Theta_2)} = \left(\frac{c_1}{c_2}\right) \angle(\Theta_1 - \Theta_2)$$

Most scientific calculators are capable of doing conversions fairly easily. Some of these calculators even handle complex number arithmetic (usually in rectangular form). Some calculators are capable of doing complex number arithmetic in mixed form.

CHAPTER 4   Impedance Circuits with AC Sources (Phasor Form)   53

In Chapter 5 we examine the response of circuits to sinusoidal inputs (time domain). In this chapter we examine the response of circuits to inputs in phasor form (frequency domain).

In *Schematics*, **VSRC** can be used as a phasor voltage source and **ISRC** can be used as a phasor current source. **VPRINT1** can be used to measure AC voltages, whereas **IPRINT** can be used to measure AC currents. In *Schematics*, both **VPRINT1** and **IPRINT** are represented by small printer symbols.

## 4.3   IMPEDANCE OF A RESISTOR

Impedance is represented by the letter Z with a bar over it ($\overline{Z}$). It represents the ratio of voltage to current in phasor form for a particular electrical device. Impedance has a magnitude and an angle and uses the units of ohms ($\Omega$). The impedance for a resistor is:

$$\overline{Z}_R = \frac{\overline{V}_R}{\overline{I}_R} = R \angle 0° \, \Omega$$

## 4.4   RESISTOR WITH AN AC VOLTAGE SOURCE

The voltage across a resistor and the current through it are always in phase. The ratio of the voltage across the resistor to the current through it (impedance) is always the value of the resistance. A 10 volt source at 0 degrees, applied across a 2k$\Omega$ resistor, would cause a 5mA current at 0 degrees.

This can be shown in *Schematics* either by using the IPRINT symbol or by examining the output in PROBE.

### A Resistor with an AC Voltage Source (Using IPRINT)

Draw the circuit diagram shown in Figure 4-1 as follows. Click **Draw/Get New Part**. Type *VSRC* and click **OK**. Place the source as shown and click the right mouse button to end placement. Double click the source. Click **DC/Change Display/Display Value/Display Name/OK**. Double click **AC**. Type *10* <space bar> *0* and click **Save Attr/Change Display/Display Value/Display Name/OK/OK**. Click **AC= 10 0** and click and drag to the right below **V1**. This shows that this

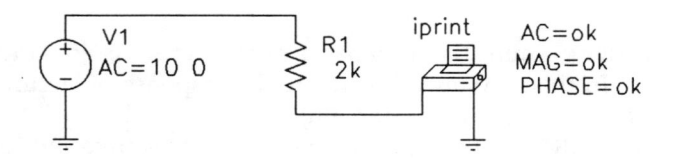

## 54 CHAPTER 4 Impedance Circuits with AC Sources (Phasor Form)

```
     V1           R1      iprint    AC=ok
    AC=10 0       2k                MAG=ok
                                    PHASE=ok
```

**4–1 Circuit for AC current through resistor**

is a 10 volt AC source at 0 degrees. Click **Draw/Get New Part**. Type *R* and click **OK**. Rotate the resistor 3 times with <Ctrl><R>. Place the resistor as shown in Figure 4–1. Click the right mouse button. Click **R1**. Click **1k**. Click and drag **1k** below R1. Double click **1k**, type *2k*, and click **OK**. Click **Draw/Get New Part**. Type *IPRINT* and click **OK**. Place the printer symbol as shown in Figure 4–1. Click the right mouse button. Double click the printer symbol. IPRINT serves as an ammeter. Double click **AC=**, type *OK*, and click **Save Attr/Change Display/Display Value/Display Name/OK**. Double click **MAG=**, type *OK*, and click **Save Attr/Change Display/Display Value/Display Name/OK**. Double click **PHASE=**, type *OK*, and click **Save Attr/Change Display/Display Value/Display Name/OK/OK**. Click and drag these attributes to the position shown in Figure 4–1. Click **Draw/Get New Part/Browse/Port/GND_EARTH/OK**. Place one ground symbol touching the lower terminal of V1 and another touching the right terminal of the ammeter labeled **IPRINT**. Click the right mouse button. Click **Draw/Wire**. Click on the upper terminal of V1, stretch the wire to the upper terminal of R1, and click. Tap the space bar. Click the lower terminal of R1, stretch the wire to the left terminal of **IPRINT**, and double click. To set up the analysis, click **Analysis/Setup.../AC Sweep/Linear**. Click in the **Total Pts.:** box, delete the contents, and type *1*. Click in the **Start Freq.:** box, delete the contents, and type *1k*. Click in the **End Freq.:** box, delete the contents, and type *1k*. This sets up a frequency analysis consisting of 1 point starting at 1kHz and ending at 1kHz. Click **OK**. Click in the **Enable** box to the left of the **AC Sweep** box to enable the AC Sweep. Click **OK**. To disable PROBE, click **Analysis/Probe Setup/Do Not AutoRun Probe/OK**. To run the analysis, click **Analysis/Run PSpice**. Type *Fig 4–1* and click **OK**. When the analysis is finished, minimize the PSpice window. To see the results, click **Analysis/Examine Output**. Scroll down until the AC ANALYSIS section appears. Here the results are shown as: FREQ: 1.000E+03 IM(V_PRINT1): 5.000E-03 IP(V_PRINT1): 0.000E+00. These results indicate that at 1kHz, the current through the ammeter labeled IPRINT has a magnitude of 5mA and a phase of 0 degrees. A hard copy of the output file may be obtained by clicking **File/Print**. Click **File/Exit** to exit NOTEPAD.

## A Resistor with an AC Voltage Source (Using PROBE)

Basically the same results may be obtained by using PROBE. Use the preceding instructions to draw the circuit diagram shown in Figure 4–2. To enable PROBE, click **Analysis/Probe Setup/Automatically Run Probe/OK**. To set up the analysis, click **Analysis/Setup/AC Sweep/Linear**. Click inside the **Total Pts.:** box, delete the contents, and type 2. Click inside the **Start Freq.:** box, delete the contents, and type *1k*. Click inside the **End Freq.:** box, delete the contents, and type *2k*. Click **OK**. Click in the box to the left of the **AC Sweep** box to enable the AC Sweep. Click **OK**. Two points had to be selected because PROBE does not run with fewer than two data points. To run the analysis, click **Analysis/Run PSpice**. Type *Fig4–2* and click **OK**. When the analysis is completed, the PROBE window opens. To obtain the magnitude data, click **Trace/Add**. Type *IM(PRINT1)* and click **OK**. To obtain the phase data, click **Plot/Add Plot/Trace/Add**. Type *IP(print1)* and click **OK**. To use **Cursors** to pinpoint values, click **Tools/Cursor/Display**. Click the left mouse button on the upper trace just to the right of the 1kHz mark. Tap the <right cursor arrow> until the cursor B1 is at exactly 1kHz. The **Probe Cursor Window** shows that at 1kHz, the phase is 0.000 degrees. Click the right mouse button on the **IM(PRINT1)** box of the lower trace. Click the right mouse button on the magnitude trace just to the right of the 1kHz mark. Hold the <shift> key depressed while tapping the <left cursor arrow> until cursor B2 is exactly on 1kHz. The **Probe Cursor Window** shows that B2 is at 1kHz and 5mA. To obtain a printout, click **File/Print/OK**.

### 4.5 IMPEDANCE OF A CAPACITOR

The opposition of a capacitor to sinusoidally alternating current is referred to as *capacitive reactance*, is represented by the abbreviation $X_c$,

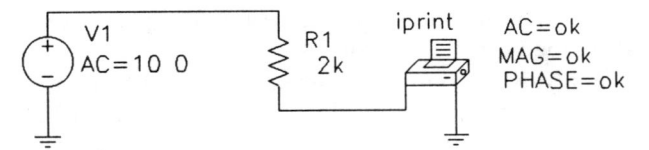

**4–2** Circuit for current through resistor using PROBE

and is measured in ohms. Capacitive reactance is inversely proportional to both the radian frequency and the capacitance value.

$$X_C = \frac{1}{\omega \cdot C}$$

In a capacitor, the current always leads the voltage by 90 degrees. The impedance of a capacitor is the ratio of the phasor voltage across the capacitor divided by the phasor current through the capacitor. It turns out that this ratio is also equal to the capacitive reactance at −90 degrees.

$$\overline{Z}_C = \frac{\overline{V}_C}{\overline{I}_C} = X_C \angle -90° \; \Omega$$

In rectangular form, capacitive impedance is:

$$\overline{Z}_C = (0 - jX_C) \; \Omega$$

A 0.1 microfarad capacitor at 1kHz would have a reactance of 1.592kΩ and an impedance of 1.592kΩ at −90 degrees in polar form (0−j1.592kΩ in rectangular form):

$$X_C = \frac{1}{2 \cdot \pi \cdot f \cdot C} = \frac{1}{6280 \cdot 0.1\mu} = 1.592 k\Omega$$

$$\overline{Z}_C = 1.592 k\Omega \angle -90° = (0 - j1.592k)\Omega$$

If a 10 volt 0 degree source is applied across the capacitor, we would expect 6.281mA at 90 degrees.

$$\overline{I}_C = \frac{\overline{V}_c}{\overline{Z}_C} = \frac{10V \angle 0°}{1.592 k\Omega \angle -90°} = 6.281 mA \angle 90°$$

## 4.6 CAPACITOR WITH AN AC VOLTAGE SOURCE

*Schematics* can be used to verify capacitive impedance. Use *Schematics* to draw the circuit diagram shown in Figure 4–3. The procedure is similar to the one described for the resistor.

Once the analysis is completed, examine the output by means of **Analysis/Examine Output**. Scroll down to the AC ANALYSIS section and confirm that the current through the capacitor is 6.283mA at 90 degrees. Close NOTEPAD with **File/Exit**.

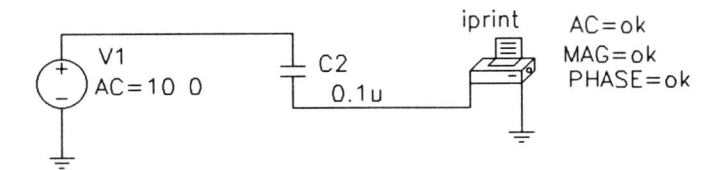

CHAPTER 4   Impedance Circuits with AC Sources (Phasor Form)   57

**4–3   Circuit for AC current through a capacitor**

## 4.7   IMPEDANCE OF AN INDUCTOR

The opposition of an inductor to sinusoidally alternating current is called *inductive reactance*, is represented by the abbreviation $X_L$, and is measured in ohms. Inductive reactance is directly proportional to both the radian frequency and the inductance value.

$$X_L = \omega L$$

In an inductor, the current always lags the voltage by 90 degrees. The impedance of an inductor is the ratio of the phasor voltage across the inductor divided by the phasor current through the inductor. It turns out that this ratio is also equal to the inductive reactance at 90 degrees.

$$\overline{Z}_L = \frac{\overline{V}_L}{\overline{I}_L} = X_L \angle 90° = 0 + jX_L \; \Omega$$

A 0.5 Henry inductor at 1kHz has an impedance of 3.14kΩ at 90 degrees.

$$\overline{Z} = \omega L \angle 90° = (6280 \cdot 0.5) \angle 90° = 3.14k\Omega \angle 90°$$

If a 10 volt 0 degree voltage source is applied across this inductor, the resulting current would be 3.185mA at –90 degrees.

$$\overline{I}_L = \frac{10V \angle 0°}{3.14k\Omega \angle 90°} = 3.185mA \angle -90°$$

## 4.8   INDUCTOR WITH AN AC VOLTAGE SOURCE

Use Schematics to draw and analyze the circuit diagram shown in Figure 4–4. The procedure is similar to the one used earlier for the resistor. R1 = 0.0001Ω must be added in series with the inductor because PSpice does not accept a pure inductor across a voltage source.

After the analysis is concluded, examine the output file and confirm

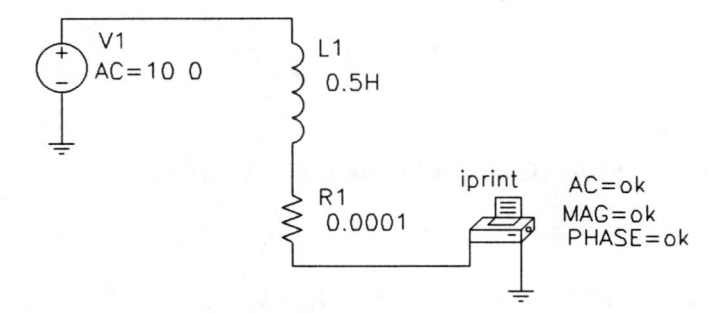

**4–4  Circuit for AC current through an inductor**

that the current through the inductor is 3.183mA at –90 degrees. Close NOTEPAD.

## 4.9  IMPEDANCES IN SERIES WITH AC VOLTAGE SOURCE

Refer to Figure 4–5. Impedances in series add. The impedance of the resistor is

$$\overline{Z}_R = (2k + j0)\ \Omega$$

while the impedance of the capacitor is

$$\overline{Z}_C = 1.592k\Omega\ \angle -90° = (0 - j1.592k)\ \Omega$$

The total impedance is the sum of the two:

$$\overline{Z}_T = \overline{Z}_R + \overline{Z}_C = (2k - j1.592k)\ \Omega = 2.56k\Omega\ \angle -38.52°$$

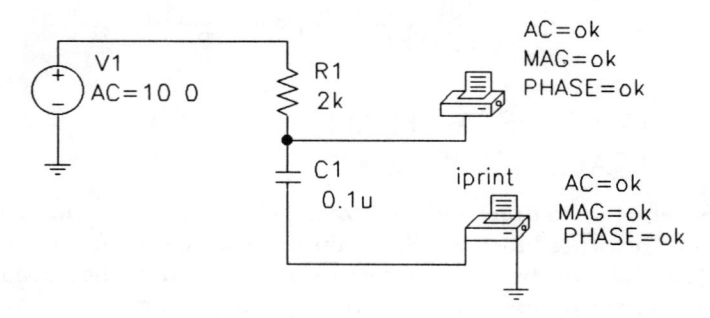

**4–5  Impedances connected in series**

CHAPTER 4   Impedance Circuits with AC Sources (Phasor Form)   59

The current may be obtained by Ohm's Law:

$$\bar{I} = \frac{\bar{V}_1}{\bar{Z}_T} = \frac{10\angle 0°}{2.56k\angle -38.52°} = 3.91mA\angle 38.52°$$

The voltage across R1 may be obtained by Ohm's Law:

$$\bar{V}_{R1} = \bar{I} \cdot \bar{Z}_R = (3.91mA\angle 38.52°)(2k\angle 0°) = 7.82V\angle 38.52° = (6.12 + j4.87)V$$

The voltage across the capacitor may be obtained by using KVL:

$$\bar{V}_C = \bar{V}_1 - \bar{V}_{R1} = (10 + j0) - (6.12 + j4.87) = (3.88-4.87)V = 6.23V\angle -51.48°$$

These results may be confirmed with *Schematics*. Use *Schematics* to draw and analyze the circuit shown in Figure 4–5. The printer symbol on top of the capacitor is an AC voltmeter. To obtain the symbol, click **Draw/Get New Part/Browse/Special/VPRINT1/OK**. Click to place the AC voltmeter as shown in Figure 4–5. Use a wire to connect the capacitor and the resistor. To use the printer symbol as an AC voltmeter, double click the symbol, double click **AC=**, type *ok*, and click **Save Attr/Change Display/Display Value/Display Name/OK**. Repeat for **MAG=** and **PHASE=**. Click **OK**. Use the same AC Sweep as in section 4.4. After the analysis is concluded, examine the output file. The output file should confirm that the magnitude of the current through the ammeter labeled IPRINT, IM(V_PRINT1), is 3.91mA and its phase, IP(V_PRINT1), is 38.52 degrees. The magnitude of the voltage at node 2, VM(2), is 6.23V while its phase, VP(2), is –51.48 degrees. Note that the voltage at node 2 is also the voltage across the capacitor.

## 4.10  IMPEDANCES IN PARALLEL WITH AC VOLTAGE SOURCE

Admittance is the reciprocal of impedance. Admittance is to impedance as conductance is to resistance. Admittance is represented by an upper case Y with a bar over it ($\bar{Y}$) and is measured in Siemens (abbreviated: S).

$$\bar{Y} = \frac{1}{\bar{Z}}$$

Admittances in parallel add. Refer to the circuit shown in Figure 4–6.

$$\bar{Y}_{R1} = \frac{1}{\bar{Z}_{R1}} = \frac{1}{2k\Omega\angle 0°} = 0.5mS\angle 0° = (0.5 + j0)mS$$

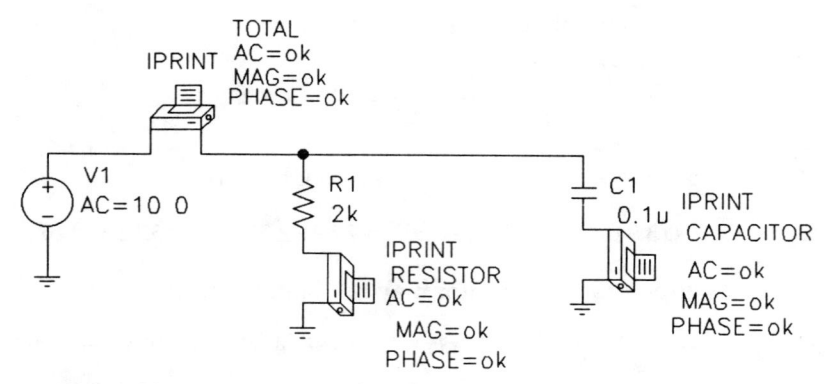

**60** CHAPTER 4 Impedance Circuits with AC Sources (Phasor Form)

$$\overline{Y}_{C1} = \frac{1}{\overline{Z}_{C1}} = \frac{1}{1.592k\Omega\angle-90°} = 0.628mS\angle 90° = (0 + j0.628)mS$$

The total admittance is the sum of the two.

$$\overline{Y}_T = \overline{Y}_{R1} + \overline{Y}_{C1} = (0.5 + j0.628)mS = 0.803mS\angle 51.474°$$

The total impedance is the reciprocal of the total admittance.

$$\overline{Z}_T = \frac{1}{\overline{Y}_T} = \frac{1}{0.803mS\angle 51.474°} = 1.245k\Omega\angle 51.474°$$

Another way of obtaining the total impedance is to use the "product over sum" formula:

$$\overline{Z}_T = \frac{\overline{Z}_{R1} \cdot \overline{Z}_{C1}}{\overline{Z}_{R1} + \overline{Z}_{C1}} = \frac{(2k\Omega\angle 0°)(1.592k\Omega\angle-90°)}{(2 - j1.592)k\Omega} = 1.245k\Omega\angle-51.474°$$

The total current may be obtained by Ohm's Law:

$$\overline{I}_T = \frac{\overline{V}_1}{\overline{Z}_T} = \frac{10V\angle 0°}{1.245k\Omega\angle-51.474°} = 8.032mA\angle 51.474° = (5.003 + j6.284)mA$$

The current through each component may also be found individually by Ohm's Law:

$$\overline{I}_{R1} = \frac{\overline{V}_1}{\overline{Z}_{R1}} = \frac{10V\angle 0°}{2k\Omega\angle 0°} = 5mA\angle 0° = (5 + j0)mA$$

CHAPTER 4   Impedance Circuits with AC Sources (Phasor Form)   61

$$\overline{I}_{C1} = \frac{\overline{V}_1}{\overline{Z}_{C1}} = \frac{10V\angle 0°}{1.592k\Omega\angle -90°} = 6.281mA\angle 90° = (0 + j6.281)mA$$

The total current may be obtained by KCL.

$$\overline{I}_T = \overline{I}_{R1} + \overline{I}_{C1} = (5 + j6.281)mA = 8.028mA\angle 51.478°$$

*Schematics* may be used to confirm these results. Use *Schematics* to draw and analyze the circuit diagram shown in Figure 4–6. To name the ammeters I-TOT-RES and I-CAP and to show the name, perform the following on each printer symbol. For the total current ammeter, double click the printer symbol, scroll down to **PKGREF=**, and double click it. Type *TOTAL*, click **Save Attr/Change Display/Display Value/OK/OK**. Repeat for *RESISTOR* and *CAPACITOR*. The other options (AC=ok, etc.) should be turned on as before.

At the completion of the analysis, examine the output file. The three currents previously obtained should check favorably with the currents obtained with *Schematics*.

## 4.11  IMPEDANCES IN SERIES-PARALLEL WITH AC VOLTAGE SOURCE

Refer to the circuit diagram shown in Figure 4–7. The impedance of R2 is negligible and is included here because PSpice does not accept pure inductors.

$$\overline{Z}_{R1} = 2k\Omega\angle 0°  \quad \overline{Z}_{C1} = 1.592k\Omega\angle 90°  \quad \overline{Z}_{L1} = 3.14k\Omega\angle 90°$$

$$\overline{Z}_X = \overline{Z}_{C1} // \overline{Z}_{L1} = \frac{(1.592k\Omega\ -90°)(3.14k\Omega\ 90°)}{(-j1.592k\Omega = (h3.14k\Omega)}$$

$$= 3.229k\Omega\angle -90° = -j3.229k\Omega$$

$$\overline{Z}_T = \overline{Z}_{R1} + \overline{Z}_X = (2 - j3.229)k\Omega = 3.798k\Omega\angle -58.228°$$

The total impedance may be obtained by using *Schematics* and "Brumgnach's 1 amp rule." Figure 4–7 shows the circuit with a 1A at 0 degrees current source. The **vprint** symbol will show the total voltage generated by the current source. Because the impedance is equal to the voltage divided by the current and the current is 1A, the total impedance in ohms is the value of the voltage. (Division by 1 does not alter the value of the numerator.) The value of the **vprint** symbol labeled Ztotal may be seen by clicking **Analysis/Examine Output** and scrolling down

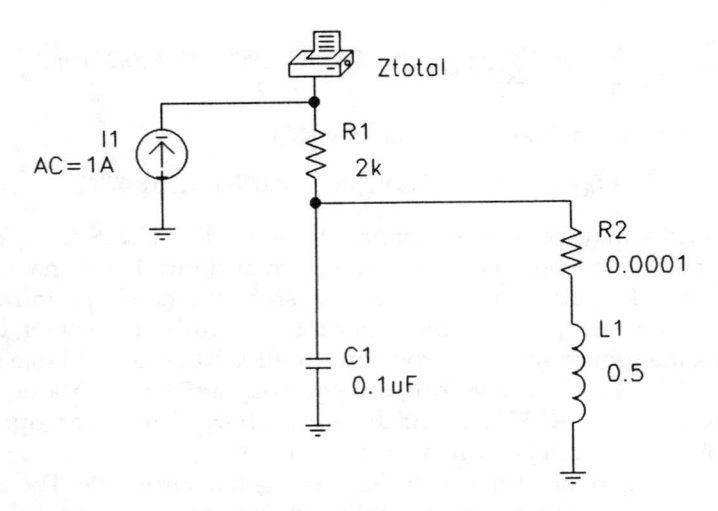

62   CHAPTER 4   Impedance Circuits with AC Sources (Phasor Form)

**4-7** Total impedance obtained with *Schematics*

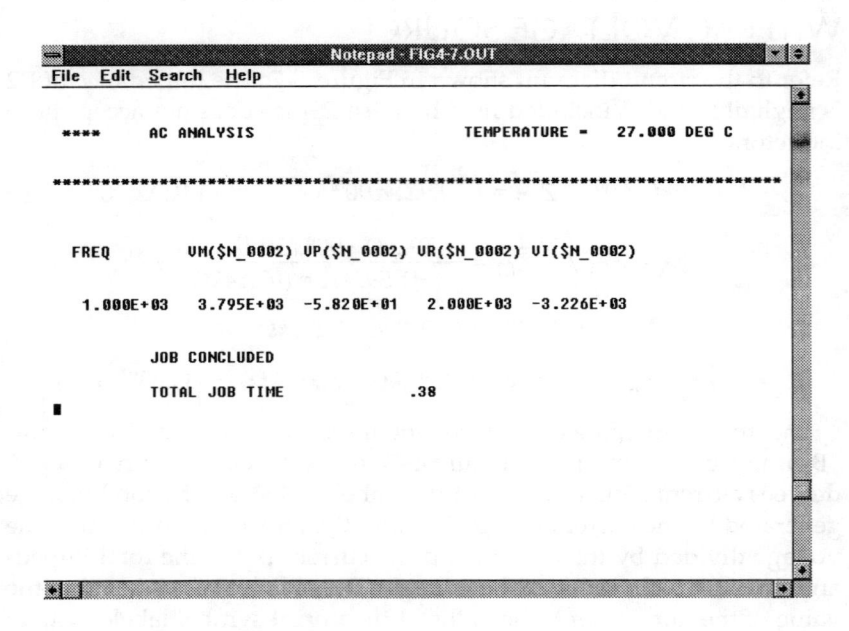

```
****    AC ANALYSIS                    TEMPERATURE =  27.000 DEG C
************************************************************************

  FREQ        VM($N_0002) VP($N_0002) VR($N_0002) VI($N_0002)

  1.000E+03   3.795E+03   -5.820E+01   2.000E+03   -3.226E+03

          JOB CONCLUDED
          TOTAL JOB TIME            .38
```

**4-8** Section of output file showing value of vprint

CHAPTER 4   Impedance Circuits with AC Sources (Phasor Form)   63

to the AC ANALYSIS section. Figure 4-8 shows the section of the output file containing the value of Ztotal.

The total current may be found by Ohm's Law:

$$\bar{I}_T = \frac{\bar{V}_1}{\bar{Z}_T} = \frac{10V\angle 0°}{3.798k\Omega\angle 58.228°} = 2.633mA\angle 58.228° = (1.386 + j2.238)mA$$

The current through the capacitor and the inductor may be found by current division:

$$\bar{I}_{C1} = \bar{I}_T \frac{\bar{Z}_{L1}}{\bar{Z}_{L1} + \bar{Z}_{C1}} = (2.633mA\angle 58.228°)\frac{3.14k\Omega\angle 90°}{(-j1.592k\Omega) + (j3.14k\Omega)}$$

$$= (2.812 + j4.54)mA = 5.34mA\angle 58.228°$$

$$\bar{I}_{L1} = \bar{I}_T \frac{\bar{Z}_{C1}}{\bar{Z}_{C1} + \bar{Z}_{L1}} = (2.633mA\angle 58.228°)\frac{1.592k\Omega\angle -90°}{(-j1.592k\Omega) + (j3.14k\Omega)}$$

$$= (-1.426 - j2.302)mA = 2.708mA\angle -121.772°$$

The total current may also be found by KCL.

$$\bar{I}_T = \bar{I}_{C1} + \bar{I}_{L1} = (2.812 + j4.54)mA + (-1.426 - j2.302)mA$$

$$= (1.386 + j2.238)mA = 2.632mA\angle 58.23°$$

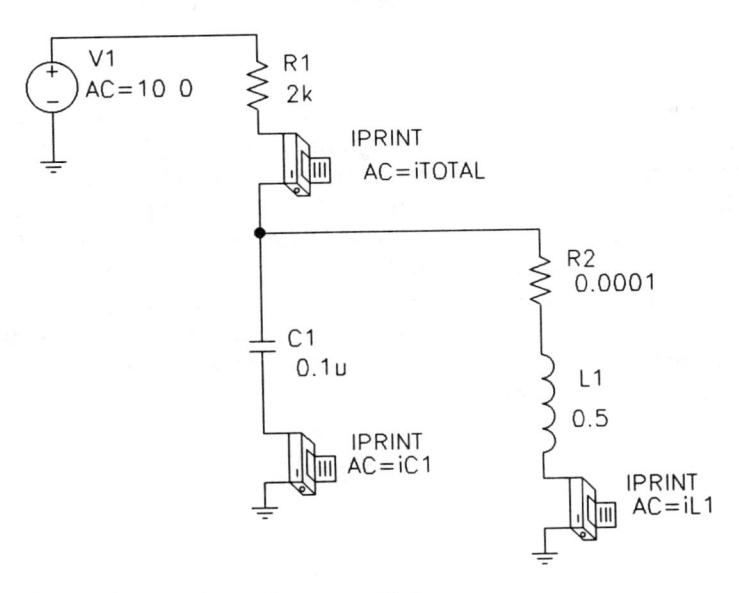

## 4-9 Impedances in series-parallel

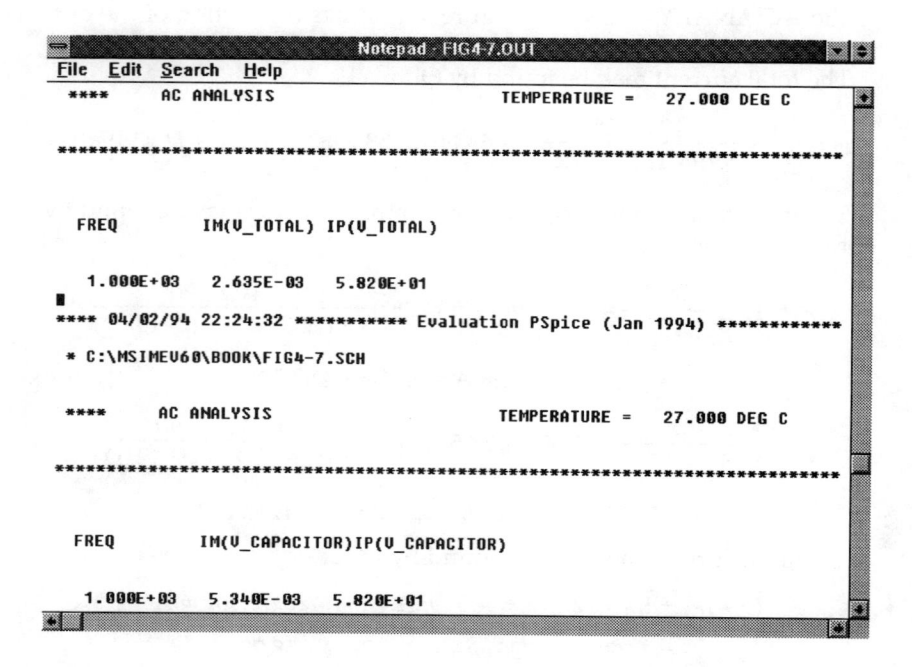

## CHAPTER 4  Impedance Circuits with AC Sources (Phasor Form)

```
****     AC ANALYSIS                    TEMPERATURE =   27.000 DEG C
***************************************************************

 FREQ         IM(V_TOTAL) IP(V_TOTAL)

 1.000E+03    2.635E-03   5.820E+01

**** 04/02/94 22:24:32 *********** Evaluation PSpice (Jan 1994) ************
 * C:\MSIMEV60\BOOK\FIG4-7.SCH

****     AC ANALYSIS                    TEMPERATURE =   27.000 DEG C
***************************************************************

 FREQ         IM(V_CAPACITOR) IP(V_CAPACITOR)

 1.000E+03    5.340E-03   5.820E+01
```

**4–10  AC portion of output file for circuit shown in Figure 4–9**

Use *Schematics* to draw and analyze the circuit shown in Figure 4-9. Change **AC=** to *iTOTAL* for the total ammeter, to *iC1* for the capacitor ammeter, and to *iL1* for the inductor ammeter. In each case activate the **Display Value** and **Display Name** check boxes. Change the **PKGREF=** designation to *iTOT* for the ammeter measuring the total current, to *iC1* for the capacitor ammeter, and to *iL1* for the inductor ammeter. Use the same AC Sweep as before. When the analysis is completed, examine the AC ANALYSIS portions of the output file and confirm that the three currents obtained with PSpice agree with the ones obtained earlier. The results are shown in Figure 4-10.

# Chapter 5   RLC CIRCUITS WITH SINUSOIDAL SOURCES

## 5.1   SINUSOIDAL VOLTAGE SOURCE (VSIN)

*Schematics* offers a sinusoidal voltage source named **VSIN**. VSIN has the following parameters, as illustrated in Figure 5–1.

voff = offset voltage
vampl = peak value of sinusoid
td = delay time
freq = frequency of sinusoid
df = damping factor
phase = phase of sinusoid

The VSIN source causes the voltage to start at $voff + vampl \cdot \sin(phase \cdot \frac{\pi}{180})$ and stay there for *td* seconds. Then the voltage becomes an exponentially damped sine wave described by

$$voff + vampl \cdot \sin\left\{2\pi \left[freq \cdot (time - td) + \frac{phase}{360}\right]\right\} \cdot e^{-(time - td) \cdot df}$$

The example in Figure 5–1 shows typical parameter values. Usually the time delay, the damping factor, and the phase are set to zero. As a matter of fact, their default value is zero. The offset voltage, peak amplitude, and frequency must be stated. *Schematics* offers a similar sinusoidal current source named **ISIN**.

## 5.2   RESISTOR WITH A SINUSOIDAL VOLTAGE SOURCE (USING PROBE)

Draw the circuit diagram shown in the right side of Figure 5–2 using *Schematics*. Use VSIN for V1 and V2. In *Schematics*, resistors, inductors,

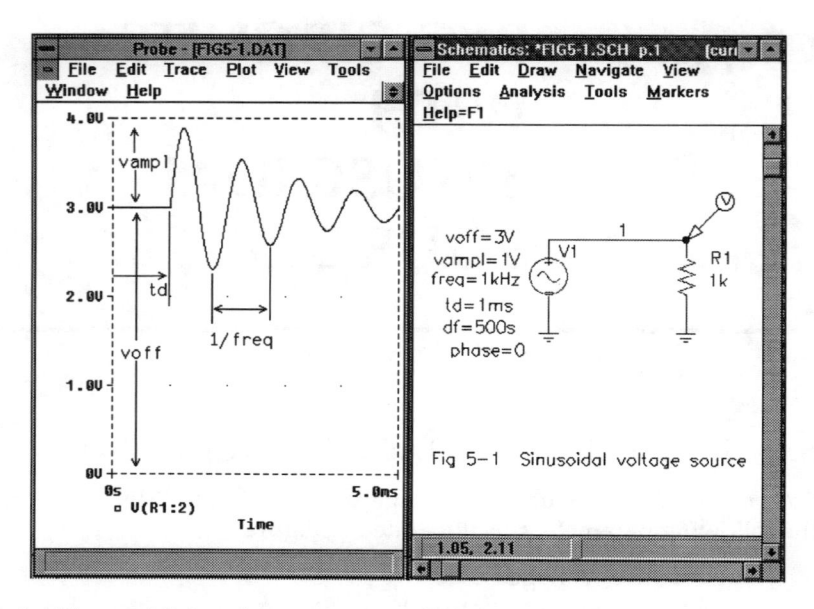

**5–1  Sinusoidal voltage source VSIN**

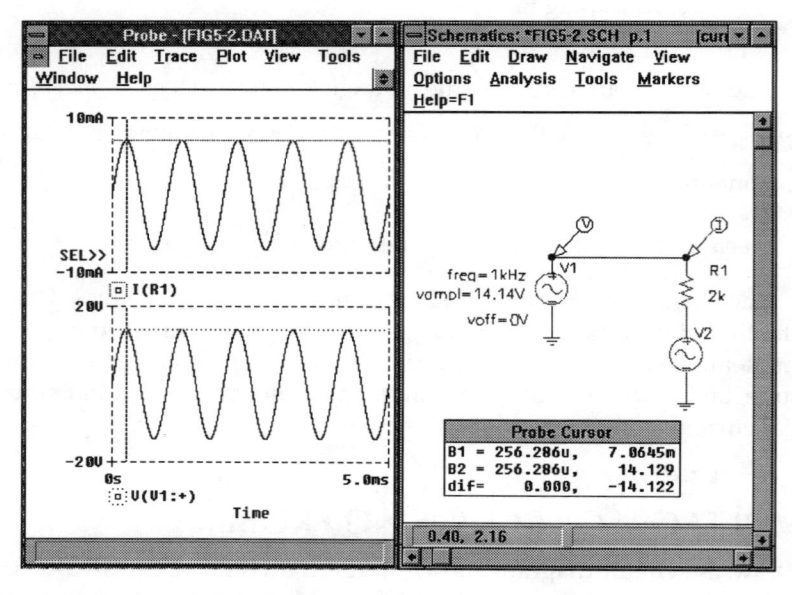

**5–2  A resistor with a sinusoidal voltage source**

CHAPTER 5   RLC Circuits with Sinusoidal Souces                           67

and capacitors have built-in ammeters. The current through one of these components can be viewed in PROBE either by identifying the current—I(R1), for example—or by placing a CURRENT MARKER at the component pin. The current through the component is considered positive if the current enters the LEFT pin of the component (when the component is first placed in the schematic diagram). Care must be taken when the component is rotated so that this "positive" current pin can still be identified after the rotation. In this example, after placing the resistor, rotate it three times. This puts the positive current pin on top. Double click V1 and assign *0V* to the voff attribute, *14.14V* to the vampl attribute, and *1kHz* to the freq attribute. For each attribute, activate display name and display value. Double click **V2** and assign a zero value for each attribute (voff, vampl, and freq). V2 will be used as an ammeter. Set up a transient analysis as follows. Click **Analysis\Setup...\Transient**. Click inside the *Print Step* dialog box, delete the contents, and type *1ns*. Click inside the *Final Time* dialog box, delete the contents, and type *5ms*. Click in the *Step Ceiling* dialog box and type *1us*. Click **OK**. Enable the **Transient...** analysis by clicking on the **Enable** box. Click **OK**. Click **Analysis/Run PSpice**. Name the circuit *Fig5–2*. After the analysis is completed, the PROBE window opens. Minimize the PSpice window if it is open. To view the input voltage, in the *Schematics* window click **Markers/Mark Voltage Level** and place the voltage level marker as shown in Figure 5–2. Click the right mouse button and two things happen at the same time. The voltage level marker becomes active (turns red) and the voltage at that point is displayed by PROBE. To view the current through the ammeter, in the PROBE window click **Plot/Add Plot** and then click **Trace/Add/I(V2)/OK**. The wave shape of the current through the ammeter and the resistor appears. The same current may be obtained as follows. In the PROBE window, click **Plot/Add Plot**. In the *Schematics* window, click **Markers/Mark Current into Pin** and place the current marker on the top pin of the resistor, as shown in Figure 5–2. Click the right mouse button. This causes the marker to become active and the resistor current to be displayed in the third plot in the PROBE window. The left side of Figure 5–2 shows the PROBE display for the applied voltage and the resulting current through the resistor. Note that the voltage and current through the resistor are in phase and that at any point in time the current may be found by applying Ohm's Law. For example, at 250µs (one-quarter cycle), the voltage is at its peak of 14.14V and the current is also at its peak 7.07mA (14.14V divided by 2kΩ). To use the CURSOR feature in PROBE, click **Tools/Cursor/Display**. To move the cursor to 250µs, click

**Tools/Cursor/Search Commands** and type *Search Forward XVAL (250μs)* and click **OK** or type *SFXVAL(250μs)* and click **OK**. The cursor information box shows the pertinent data. The procedure may be repeated for the second trace. Click the right mouse button on the origin of the other trace. Click **Tools/Cursor/Search Commands,** type *SFX VAL(250μs) and click OK. The cursor information box shows the additional data. To set up the Schematics editor for the next circuit diagram, click* **File/New/Yes.**

## 5.3 CAPACITOR WITH A SINUSOIDAL VOLTAGE (USING PROBE)

In sections 4.5 and 4.6 we showed that the current through a capacitor leads the voltage across a capacitor by 90 degrees. Thus, if the voltage across a capacitor is a sine, the current through it must be a cosine. *Schematics* may be used to illustrate this fact by using the TRANSIENT analysis feature.

The current and voltage for a capacitor are related by the equation:

$$i = C\frac{dv}{dt}$$

If the voltage is a sine of the form:

$$v = V_{peak}\sin(\omega t)$$

the current then becomes:

$$i = C\frac{d[V_{peak}\sin(\omega t)]}{dt} = \omega C V_{peak}\cos(\omega t) = \omega C V_{peak}\sin(\omega t + 90°)$$

Here we can recognize again that because the voltage is multiplied by $\omega C$ to obtain the current, then $\frac{1}{\omega C}$ must be the opposition of a capacitor to alternating current. This opposition is referred to as *capacitive reactance* ($X_C$).

For the circuit shown in Figure 5-3, use the procedure outlined in section 5.2 to draw, analyze, and display the wave forms in PROBE. Remember to rotate the capacitor three times.

From Figure 5-3 it should be apparent that in a capacitor the current leads the voltage by 90 degrees and that the capacitive reactance is $\frac{1}{\omega C}$. In sections 4.5 and 4.6 we applied a phasor voltage of 10V at 0 degrees. The 10V is the RMS value of a 14.14V peak sinusoid. In this example we

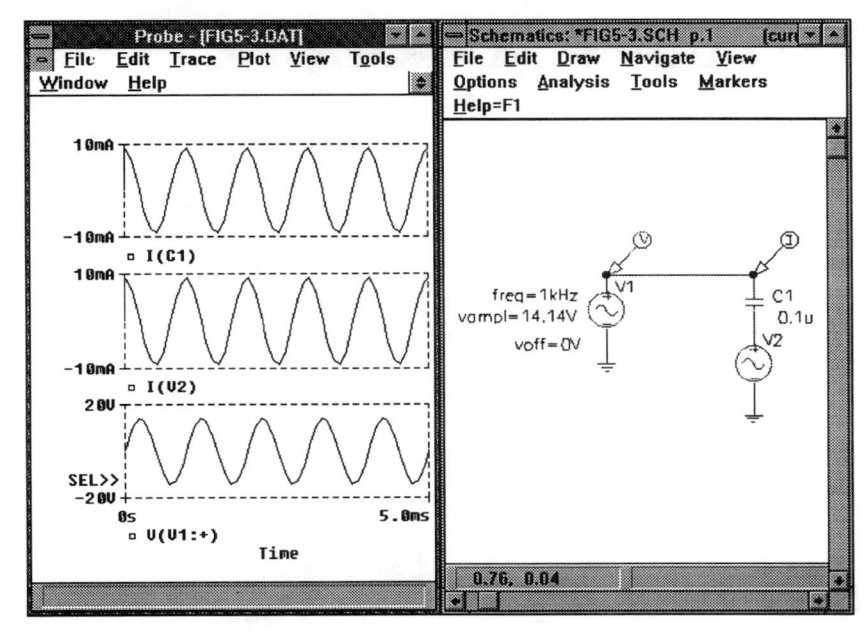

CHAPTER 5    RLC Circuits with Sinusoidal Souces                              69

**5–3  A capacitor with a sinusoidal voltage source**

used the peak value of 14.14V. The resulting peak current may be found by Ohm's Law as:

$$I_{peak} = \frac{V_{peak}}{X_C} = \frac{14.14V}{1.592k\Omega} = 8.88mA$$

Figure 5–4 shows a closer examination of the capacitor current, with the CURSOR indicating the peak current.

To use the CURSOR feature in PROBE, proceed as follows. Use **Trace/Add/I(V2)/OK** to display the current through V2. Use **Tools/Cursor/ Display** to display the CURSOR feature. Use **Tools/Cursor/Max** to find the peak value of the current. The CURSOR dialog box shows the peak value of the current and the CURSOR cross-line shows where the peak is. It should be clear that the results obtained with PSpice agree very closely with the results obtained from the formulas. The reason that they are not exactly the same is that the analysis in PSPICE was not done in small enough steps (as in the previous section involving the

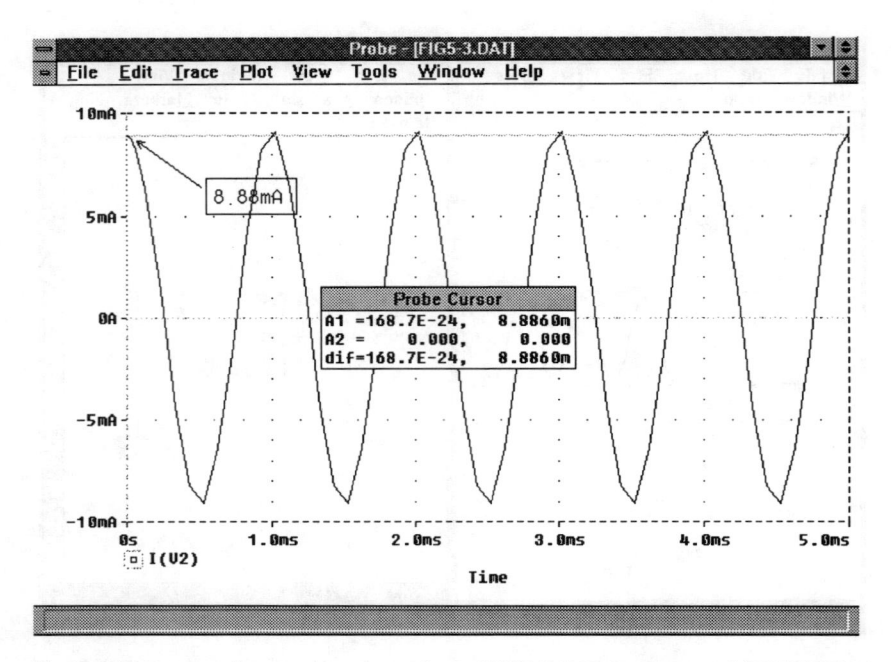

**5–4** The capacitor current using CURSOR feature of PROBE

resistor). To set up the *Schematics* editor for the next circuit diagram, click **File/New/Yes**.

## 5.4 INDUCTOR WITH A SINUSOIDAL CURRENT SOURCE (USING PROBE)

In sections 4.7 and 4.8 we showed that the current through an inductor lags the applied voltage by 90 degrees. Thus, if the voltage across the inductor is a sine, the current is a sine that starts 90 degrees later. *Schematics* may be used to illustrate this fact by using the TRANSIENT analysis feature.

The current and voltage for an inductor are related by the equation:

$$v = L\frac{di}{dt} = L\frac{d[I_{peak}\sin(\omega t)]}{dt} = \omega L \cdot I_{peak}\cos(\omega t) = X_L \cdot I_{peak}\sin(\omega t + 90°)$$

Here we can recognize that because the current is multiplied by $\omega L$ to obtain the voltage, then $\omega L$ must be the opposition of the inductor to

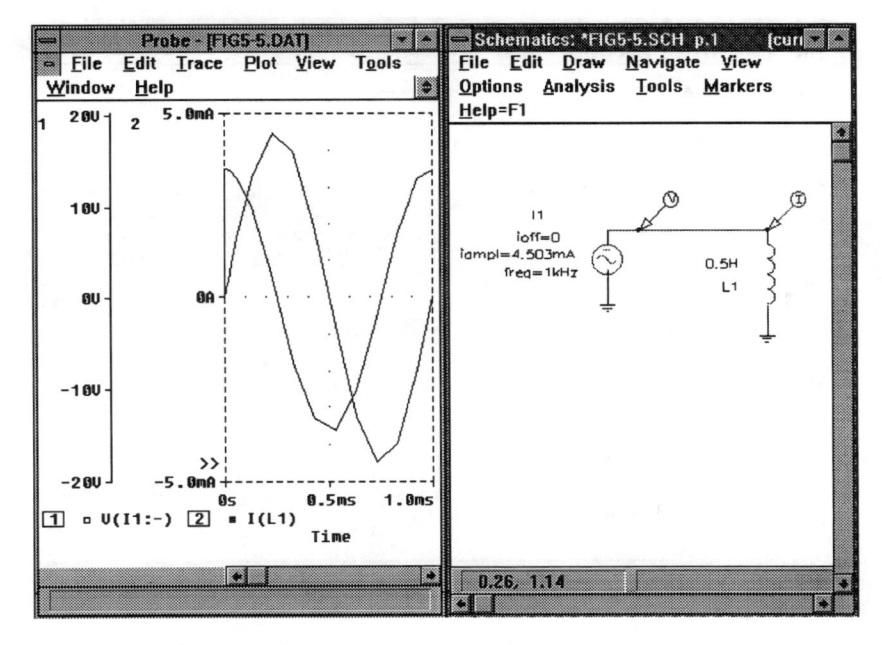

CHAPTER 5     RLC Circuits with Sinusoidal Souces                               71

## 5-5   An inductor with a sinusoidal current source

alternating current. This opposition is referred to as *inductive reactance* ($X_L$). It should also be apparent that the voltage leads the current by 90 degrees.

For the circuit shown in Figure 5-5, use the procedure outlined in section 5.2 to draw, analyze, and display the wave forms in PROBE. Use ISIN for current source I1. Note that the current source I1 is placed with its negative polarity up. PSpice current sources always cause conventional current out of the negative terminal. Also remember to rotate the inductor three times.

From Figure 5-5 it should be apparent that the current through an inductor lags the voltage by 90 degrees and that the inductive reactance is $\omega L$. In sections 4.7 and 4.8 we applied a phasor voltage of 10V at 0 degrees. The 10V is the RMS value of a 14.14V peak voltage. The resulting current was 3.185mA RMS, which is 4.503mA peak. In this example we used the peak current of 4.503mA. The resulting peak voltage may be found by Ohm's Law as:

$$V_{peak} = (I_{peak}) \cdot (X_L) = (4.503mA) \cdot (3.14k\Omega) = 14.4V$$

# CHAPTER 5   RLC Circuits with Sinusoidal Souces

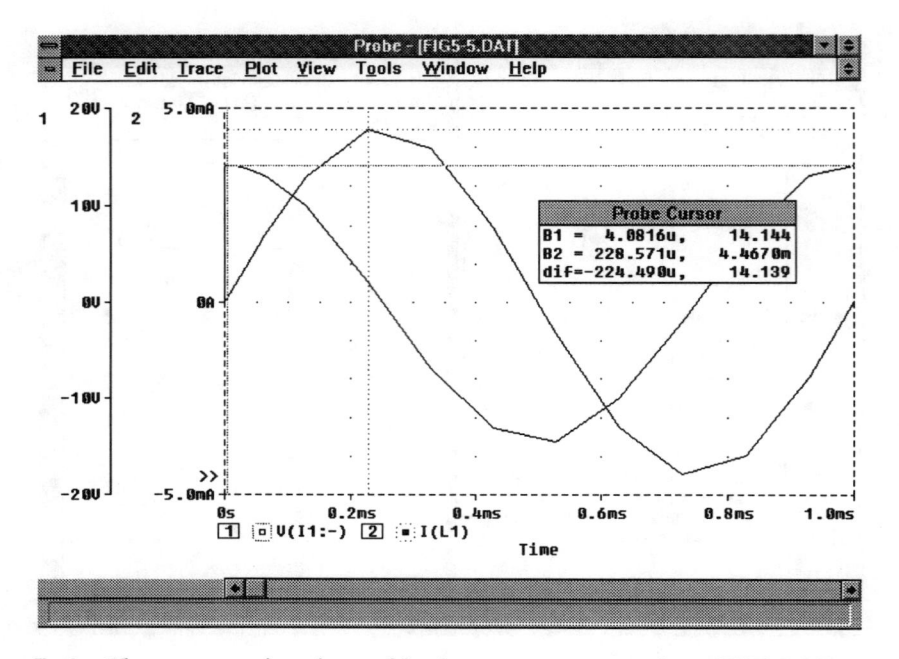

**5-6   Closer examination of inductor current using CURSOR feature of PROBE**

To obtain the PROBE display shown in Figure 5-5 proceed as follows. To display the trace of the inductor voltage, click **Markers/Mark Voltage Level**, place the voltage marker as shown in Figure 5-5, and click the right mouse button. Use **Plot/X Axis Settings/User Defined** and click inside the *To* dialog box. Delete its contents, type *1ms*, and click **OK**. This makes the new x axis settings 0 to 1 ms. To display the current through the inductor on a second axis, click **Plot/Add Y Axis**; in the *Schematics* window, click **Markers/Mark Current into Pin**, place the current marker as shown in Figure 5-5 and click the right mouse button.

Figure 5-6 shows a closer examination of the inductor current with CURSOR B2 indicating the peak current of 4.503mA and CURSOR B1 indicating the voltage at zero inductor current (approximately 14.14V). The annotations were inserted with **Tools/Label/Text**, **Tools/Label/Box**, and **Tools/Label/Arrow** (see section 2.3 on the use of these tools). To set up the *Schematics* editor for the next circuit diagram, click **File/New/Yes**.

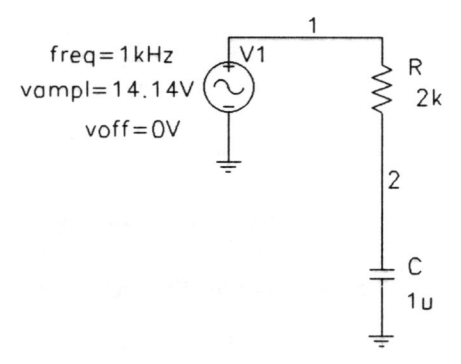

CHAPTER 5    RLC Circuits with Sinusoidal Souces    73

```
freq=1kHz
vampl=14.14V
voff=0V
```

**5-7  Series RC circuit with sinusoidal voltage source**

## 5.5  SERIES RC CIRCUIT WITH A SINUSOIDAL VOLTAGE SOURCE (USING PROBE)

The circuit in Figure 5-7 shows a resistor R connected in series with a capacitor C. The series combination is energized by a sinusoidal voltage source. The circuit current and voltage responses are each made up of a natural component and a forced component. The combination of these two components is referred to as the *total response*.

The sinusoidal voltage source has the following form:

$$v_1 = V_{peak}\sin(\omega t)$$

The total voltage response across the capacitor may be obtained by analyzing the circuit by one of many means (La Place Transforms, for example). The total voltage across the capacitor is made up of a natural component and a forced component and has the following form:

$$v_c = Ae^{-\frac{t}{RC}} + 2|B|\sin(\omega t + \varphi)$$

The first term is the natural component and the second term is the forced component. The value of A, the magnitude of B, and the value of φ may be obtained from the following:

$$A = \frac{V_{peak} \cdot \omega}{RC} \cdot \frac{1}{\left(\frac{1}{RC}\right)^2 + \omega^2}$$

$$|B| = \frac{V_{peak}}{2\sqrt{(\omega RC)^2 + 1}}$$

$$\varphi = 90° - \tan^{-1}\left(\frac{1}{-\omega RC}\right)$$

For the values shown in Figure 5–6, the capacitor voltage is:

$$v_c = 1.12e^{-\frac{t}{0.2ms}} + 1.12\sin(6280t - 85.45°) \text{ volts}$$

*Schematics* may be used to easily obtain the graph of this response. Use *Schematics* to draw the circuit diagram shown in Figure 5–7. Set up a transient analysis (see section 5.2) with a 1ns print step and a 10ms final time. Enable the transient analysis and run PSpice. In PROBE, add the trace V(2) to view the capacitor voltage. Add another plot and add trace V(1,2) to view the voltage across the resistor. Add a third plot and add trace V(1) to view the total applied voltage. At this point, check KVL by adding the trace V(1,2) + (V2) and noticing that it superimposes exactly on top of the V(1) trace. This confirms that the voltage across the resistor and the voltage across the capacitor add up to yield the total applied voltage. The resulting wave forms are shown in Figure 5–8. It should also be noted in Figure 5–8 that the transient response dies down in 5 time constants (10ms), as predicted by the preceding equation.

To check the capacitor voltage a little more closely, delete all the plots except the bottom plot showing V(2). Plots are deleted with **Plot/Delete Plot**. Activate the cursor with **Tools/Cursor/Display** and use **Tools/Cursor/Peak** to position the cursor at the first peak. The cursor coordinates show that the capacitor voltage peaks at 1.9129V and that this first peak occurs at 466.282μs. If this value of time is substituted into the total capacitor voltage response equation, the following result is obtained:

$$v_c = 1.12e^{-\frac{466.282\mu s}{0.2ms}} + 1.12\sin[(6280) \cdot (466.282) \cdot (10^{-6}) \cdot (57.3) - 85.45°]$$

$$= 1.97 \text{ volts}$$

There is obviously a discrepancy between the hand solution of 1.97V and the PSpice solution of 1.91V. The problem is inherent in the PSpice solution because the plot of the capacitor voltage shown by PROBE is too coarse. To obtain better results, go back to the analysis setup and add a step ceiling of 1μs. Run the PSpice analysis again. The analysis takes much longer, but the results are more accurate. Check the results in PROBE with **Trace/AddTrace/V(2)**. To check the value of the capaci-

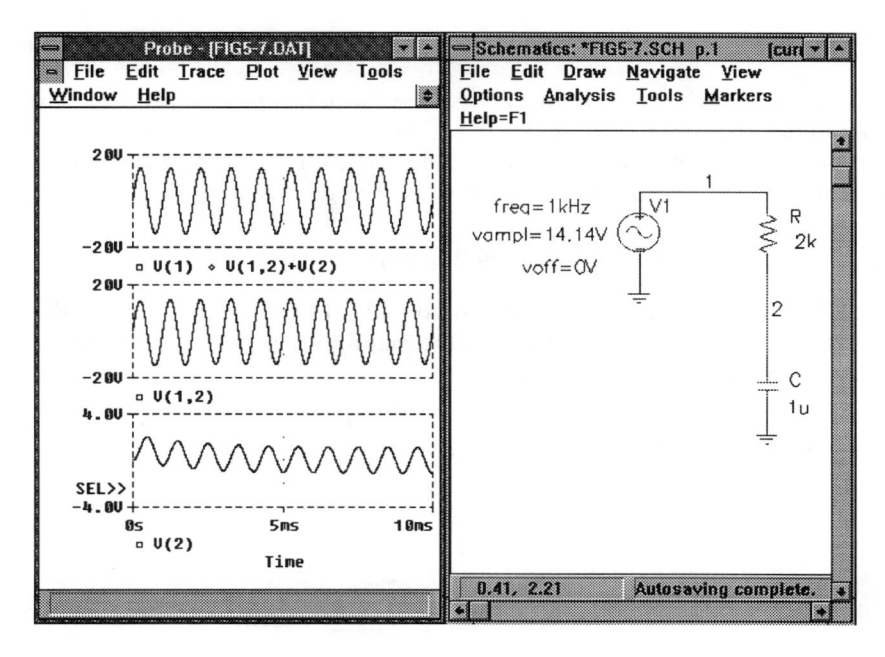

CHAPTER 5   RLC Circuits with Sinusoidal Souces   75

**5–8   Series RC response**

tor voltage and the time of the first peak, use **Tools/Cursor/Display/Tools/Cursor/Peak**. The resulting screen, shown in Figure 5–9, indicates that the first peak of the capacitor voltage is 2.0002V and occurs at 477.74µs. If this time is substituted into the equation for the capacitor response, the resulting capacitor voltage is 1.9999V. The two answers agree closely enough.

## 5.6   SERIES RL CIRCUIT WITH A SINUSOIDAL VOLTAGE SOURCE (USING PROBE)

The circuit in Figure 5–10 shows a resistor R connected in series with an inductor L. The series combination is energized by a sinusoidal voltage source. The circuit current and voltage responses are each made up of a natural component and a forced component. The combination of these two components is referred to as the *total response*.

The sinusoidal voltage source has the following form:

$$v_1 = V_{peak}\sin(\omega t)$$

The total voltage response across the inductor may be obtained by analyzing the circuit by one of many means (La Place Transforms, for example). The total voltage across the inductor is made up of a natural component and a forced component and has the following form:

$$v_L = Ae^{-\frac{R \cdot t}{L}} + 2|B|\sin(\omega t + \varphi)$$

The first term is the natural component and the second term is the forced component. The value of A, the magnitude of B, and the value of φ may be obtained as follows:

$$A = \frac{V_{peak} \cdot \omega\left(-\dfrac{R}{L}\right)}{\left(\dfrac{R}{L}\right)^2 + \omega^2}$$

$$|B| = \frac{V_{peak} \cdot \omega}{2\sqrt{\omega^2 + \left(\dfrac{R}{L}\right)^2}}$$

$$\varphi = 90° - \tan^{-1}\left(\dfrac{\dfrac{R}{L}}{-\omega}\right)$$

For the values shown in Figure 5–9, the inductor voltage is:

$$v_L = -6.41e^{-4000t} + 11.92\sin(6280t - 32.49°) \; volts$$

*Schematics* may be used to easily obtain the graph of this response. Use *Schematics* to draw the circuit diagram shown in Figure 5–10. Set up a transient analysis (see section 5.2) with a 1ns print step and a 5ms final time. Enable the transient analysis and run PSpice. In PROBE, add the trace V(2) to view the inductor voltage. Add another plot and add trace V(1,2) to view the voltage across the resistor. Add a third plot and add trace V(1) to view the total applied voltage. At this point, check KVL by adding the trace V(1,2) + (V2) and noticing that it superimposes exactly on top of the V(1) trace. This confirms that the voltage across the resistor and the voltage across the inductor add up to yield the total applied voltage. The resulting wave forms are shown in Figure 5–11. It should also be noted in Figure 5–11 that the transient response dies down in 5 time constants (100μs), as predicted by the preceding equation.

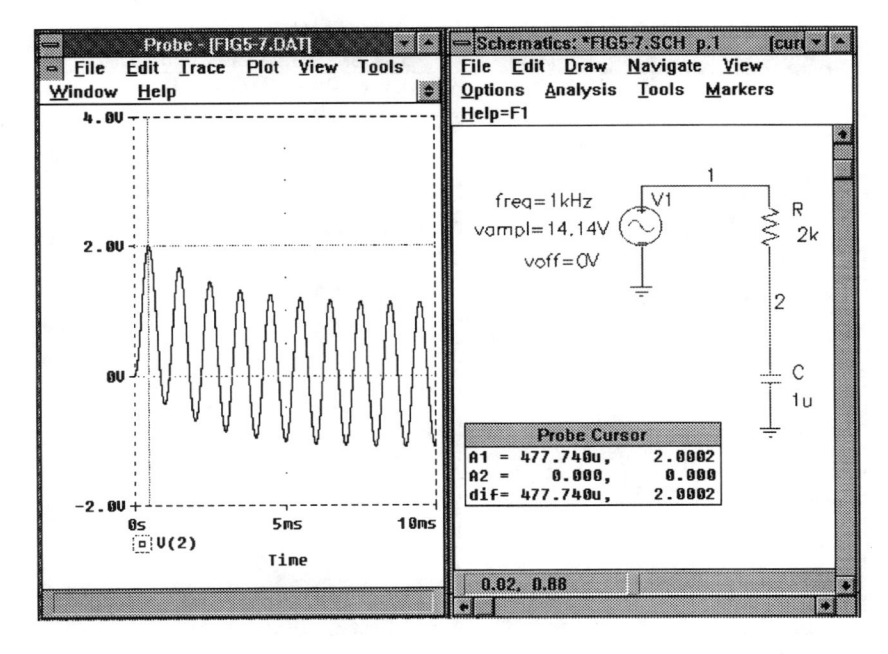

CHAPTER 5   RLC Circuits with Sinusoidal Souces   77

**5-9   A closer examination of the capacitor voltage**

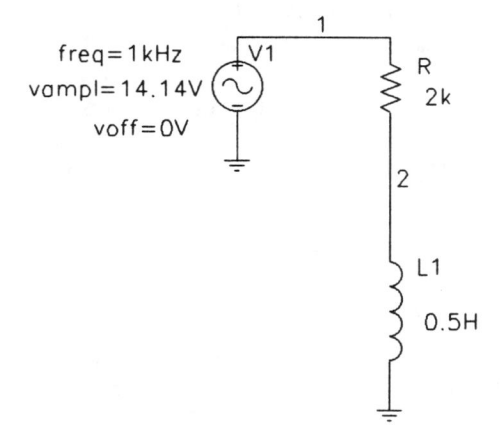

**5-10   Series RL with sinusoidal voltage source**

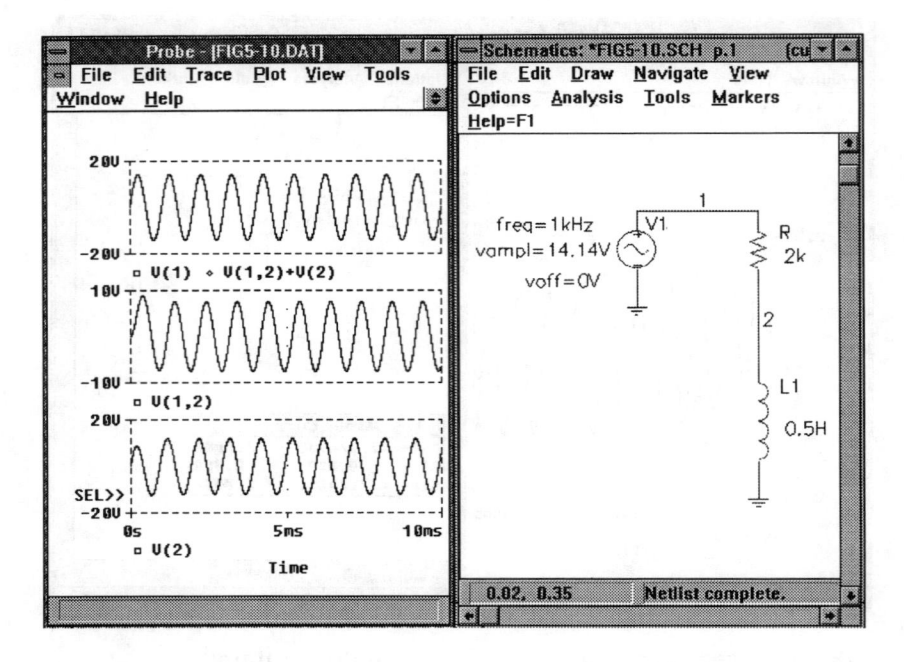

## 5–11 Series RL response

To check the inductor voltage a little more closely, delete all the plots except the bottom plot showing V(2). Plots are deleted with **Plot/Delete Plot**. Activate the cursor with **Tools/Cursor/Display** and use **Tools/Cursor/Peak** to position the cursor at the first peak. The cursor coordinates show that the inductor voltage peaks at 8.7868V and that this first peak occurs at 173.401μs. If this value of time is substituted into the total inductor voltage response equation, the following result is obtained:

$$v_L = -6.41e^{-(4000) \cdot (173.401) \cdot (10^{-6})}$$

$$+ 11.92\sin[(6280) \cdot (173.401) \cdot (10^{-6}) \cdot (57.3) - 85.45°] = 8.6731 \text{ volts}$$

There is obviously a discrepancy between the hand solution of 8.6731V and the PSpice solution of 8.7868V. The problem is inherent in the PSpice solution because the plot of the capacitor voltage shown by PROBE is too coarse. To obtain better results, go back to the analysis setup and add a step ceiling of 1μs. Run the PSpice analysis again. The

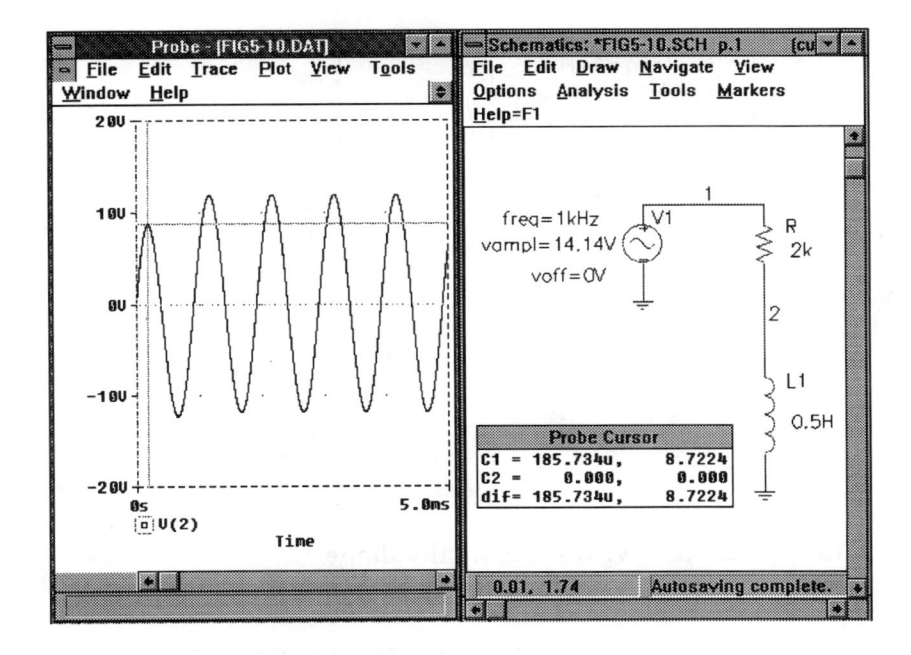

CHAPTER 5    RLC Circuits with Sinusoidal Souces                                79

**5-12  A closer examination of the inductor voltage**

analysis takes much longer, but the results are more accurate. Check the results in PROBE with **Trace/AddTrace/V(2)**. To check the value of the inductor voltage and the time of the first peak, use **Tools/Cursor/Display/Tools/Cursor/Peak**. The resulting screen, shown in Figure 5-12, indicates that the first peak of the inductor voltage is 8.7224V and occurs at 185.734µs. If this time is substituted into the equation for the capacitor response, the resulting capacitor voltage is 8.7131V. The two answers agree closely enough.

# 5.7  SERIES RLC CIRCUIT WITH A SINUSOIDAL VOLTAGE SOURCE (USING PROBE)

The circuit in Figure 5-13 shows a resistor R, an inductor L, and a capacitor C connected in series. The series combination is energized by a sinusoidal voltage source. The circuit current and voltage responses are

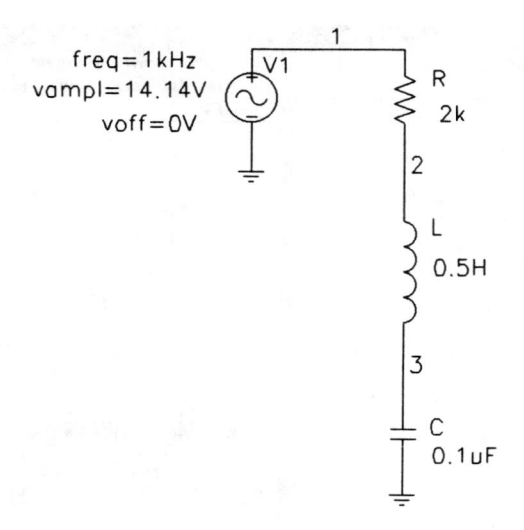

```
                    1
freq=1kHz    V1
vampl=14.14V         R
voff=0V              2k

                     2

                     L
                     0.5H

                     3

                     C
                     0.1uF
```

## 5-13 Series RLC with sinusoidal voltage

each made up of a natural component and a forced component. The combination of these two components is referred to as the *total response*.

The sinusoidal voltage source has the following form:

$$v_1 = V_{peak}\sin(\omega t)$$

The total voltage response across the inductor may be obtained by analyzing the circuit by one of many means (La Place Transforms, for example). The total voltage across the capacitor is made up of a natural component and a forced component. For the values shown in Figure 5-13, the capacitor voltage is:

$$v_c = 14.02e^{-2000t}\sin(4000t + 30.25°) + 8.92\sin(6280t + 232.27°) \; volts$$

The first term is the natural component and the second term is the forced component.

*Schematics* may be used to easily obtain the graph of this response. Use *Schematics* to draw the circuit diagram shown in Figure 5-13. Set up a transient analysis (see section 5.2) with a 1ns print step and a 5ms final time. Enable the transient analysis and run PSpice. In PROBE, add the trace V(3) to view the capacitor voltage. Add another plot and add trace V(1,2) to view the voltage across the resistor. Add a third plot and add trace V(2,3) to view the capacitor voltage. Add a fourth plot and add trace V(1) to view the total applied voltage. At this point, check KVL by

# CHAPTER 5  RLC Circuits with Sinusoidal Souces

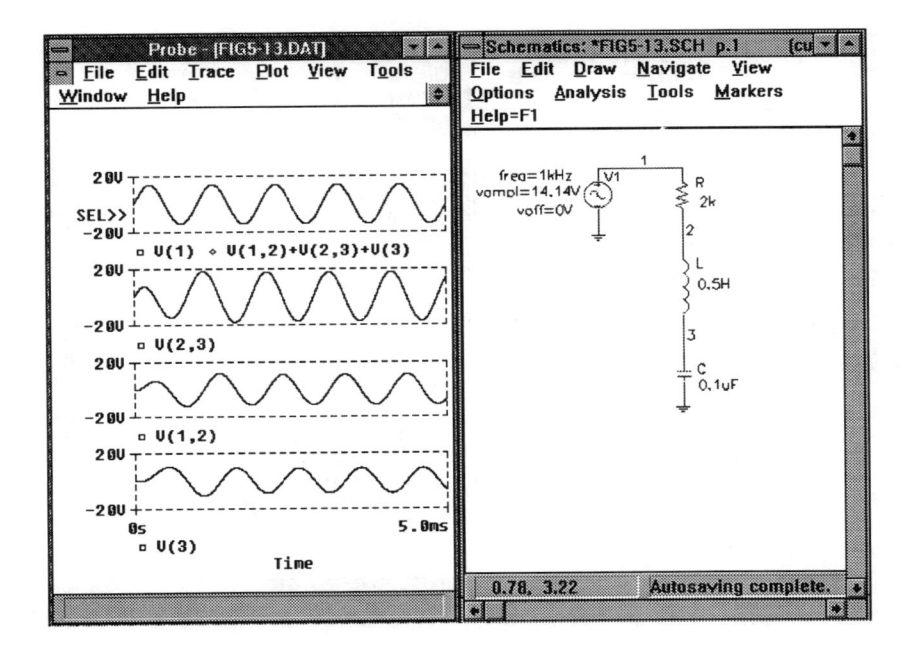

## 5-14 Series RLC response

adding the trace V(1,2) + V(2,3) + V(3) and noticing that it superimposes exactly on top of the V(1) trace. This confirms that the voltage across the resistor, the voltage across the inductor, and the voltage across the capacitor add up to yield the total applied voltage. The resulting wave forms are shown in Figure 5–14.

To check the capacitor voltage a little more closely, delete all the plots except the bottom plot showing V(3). Plots are deleted with **Plot/Delete Plot**. Activate the cursor with **Tools/Cursor/Display** and use **Tools/Cursor/Peak** to position the cursor at the first peak. The cursor coordinates show that the capacitor voltage peaks at 9.723V and that this first peak occurs at 574.781µs. If this value of time is substituted into the total capacitor voltage response equation, the following result is obtained:

$$v_c = 14.02 e^{-(2000 \cdot 574.781 \cdot 10^{-6})} \sin(4000 \cdot 574.781 \cdot 10^{-6} \cdot 57.3 + 30.25°)$$

$$+ 8.92 \sin(6280 \cdot 574.781 \cdot 10^{-6} \cdot 57.3 + 232.27°)$$

$$= 10.1323 \; volts$$

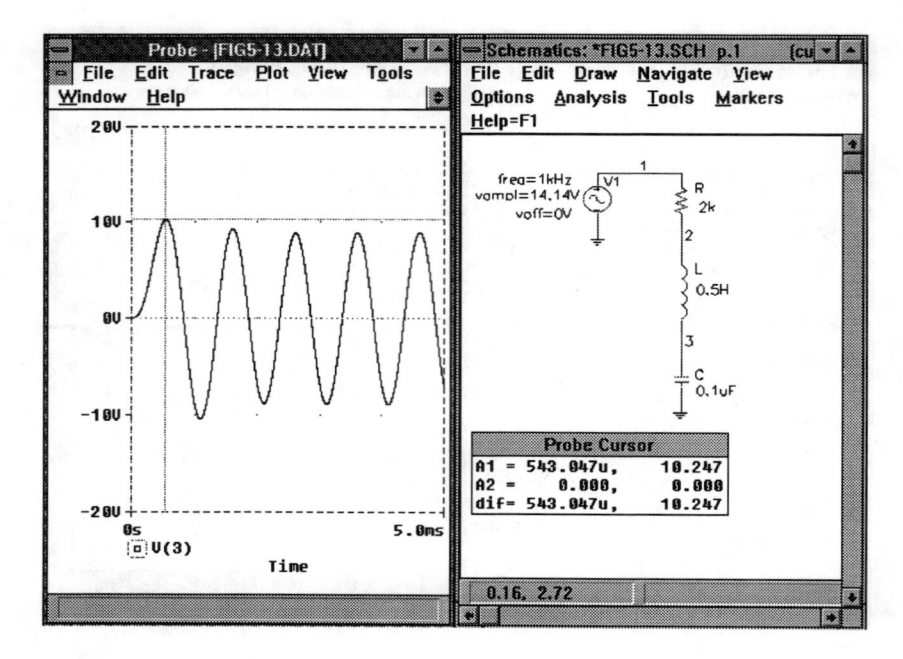

**5–15  A closer examination of the capacitor voltage for a series RLC**

There is obviously a discrepancy between the hand solution of 10.1323V and the PSpice solution of 9.723V. The problem is inherent in the PSpice solution because the plot of the capacitor voltage shown by PROBE is too coarse. To obtain better results, go back to the analysis setup and add a step ceiling of 1μs. Run the PSpice analysis again. The analysis takes much longer, but the results are more accurate. Check the results in PROBE with **Trace/AddTrace/V(3)**. To check the value of the capacitor voltage and the time of the first peak, use **Tools/Cursor/Display/Tools/Cursor/Peak**. The resulting screen, shown in Figure 5–15, indicates that the first peak of the inductor voltage is 10.247V and occurs at 543.047μs. If this value time is substituted into the equation for the capacitor response, the resulting capacitor voltage is 10.2729V. The two answers agree closely enough.

## Chapter 6 RLC CIRCUITS WITH PULSE VOLTAGE SOURCE (VPULSE)

## 6.1 PULSE VOLTAGE SOURCE (VPULSE)

*Schematics* offers a pulse voltage source named **VPULSE**. VPULSE has the following parameters, as illustrated in Figure 6–1.

| parameter | meaning | default units | default value | value in Figure 6-1 |
|---|---|---|---|---|
| v1 | low voltage | volts | must be specified | 0V |
| v2 | high voltage | volts | must be specified | 5V |
| td | initial time delay | seconds | 0 | 1ms |
| tr | rise time | seconds | print step | 0.3ms |
| tf | fall time | seconds | print step | 0.7ms |
| pw | pulse width | seconds | final time | 2ms |
| per | period | seconds | final time | 5ms |

The **print step** and **final time** values are obtained as default values from the specifications entered in the **transient analysis setup**.

## 6.2 RC CIRCUIT WITH VPULSE

A capacitor may be charged from 0 volts to a final voltage $V_F$ in a certain period of time. The equation representing this voltage buildup is:

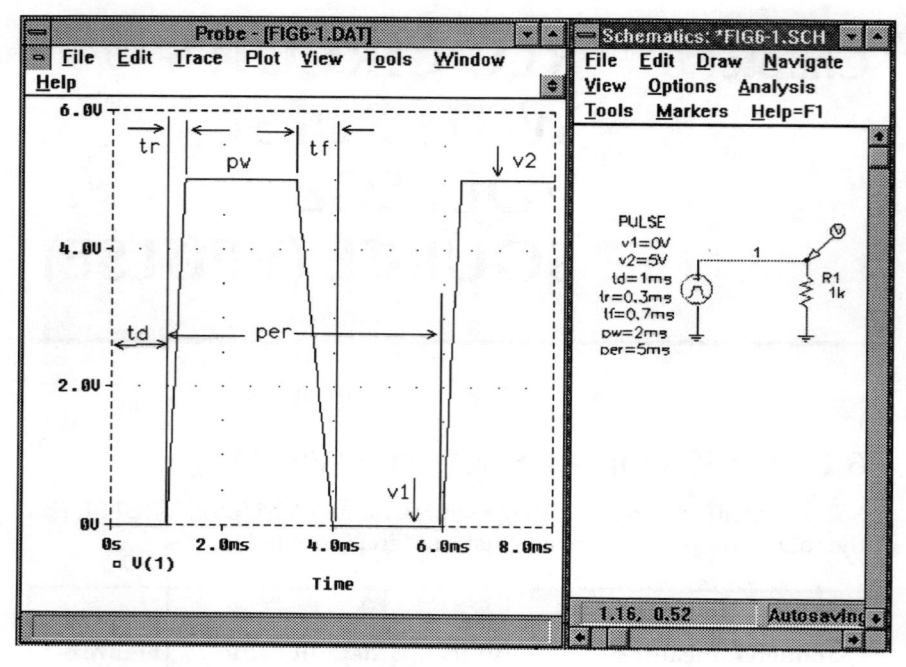

## CHAPTER 6  RLC Circuits with Pulse Voltage Source (VPULSE)

**6-1  Pulse voltage source VPULSE**

$$v_C = V_F\left(1 - e^{-\frac{t}{\tau}}\right)$$

$v_c$ = capacitor voltage at any particular time (V)
$V_F$ = final capacitor voltage (V)
e = 2.7183 (base of natural logs)
$\tau$ = RC time constant (s)
t = particular time at which $v_c$ is desired (s)

   The time in seconds that it takes for a capacitor to charge to 63.2% of its final value is referred to as the *charging time constant* and is obtained by multiplying the charging resistor value by the capacitor value.
   Capacitor charging may be easily investigated with the aid of PSpice for Windows. Figure 6-2 shows a 1µF capacitor being charged through a 1kΩ resistor by a 0 to 5 V pulse voltage source. The charging time constant is:

$$RC = (1k\Omega) \cdot (1\mu F) = 1ms$$

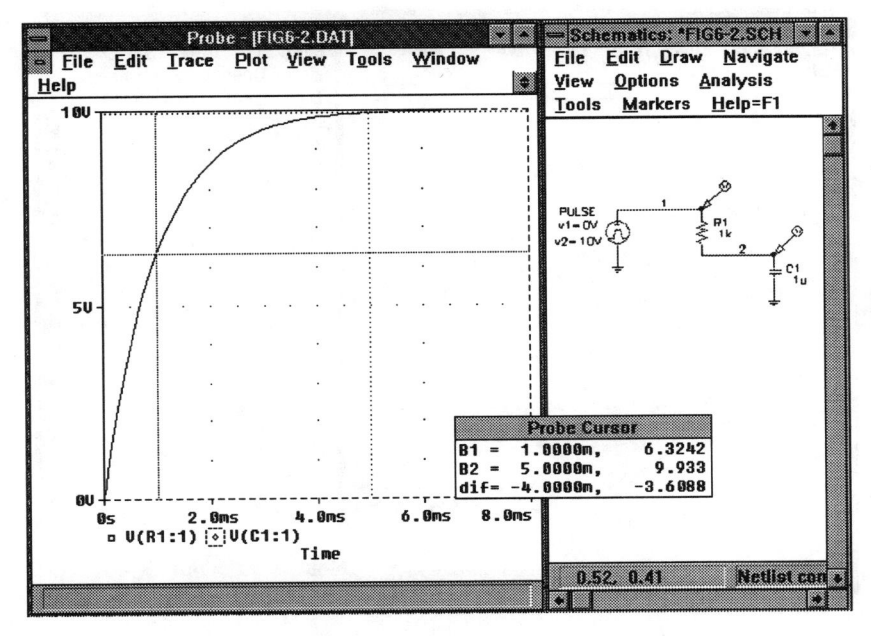

CHAPTER 6    RLC Circuits with Pulse Voltage Source (VPULSE)    85

### 6-2 Capacitor charging using VPULSE

This means that in 1ms the capacitor voltage should be at 63.2% of 5 volts (3.16V), and the maximum of 5V should be reached in 5 time constants or 5ms. Use *Schematics* to draw the circuit shown in Figure 6-2. Set up a **transient analysis** for a 1ns **print step** and a 7ms **stop time**. Set up the pulse voltage source VPULSE for v1 = 0V and v2 = 5V. Leave the other specifications at their default values. Run PSpice. Once the analysis is completed and PROBE is showing the wave forms at nodes 1 and 2, check the value of the capacitor voltage at 1ms and 5ms. This may be done as follows. Click **Tools/Cursor/Display** and click the marker to the left of the **V(C1:1)** label under the horizontal axis. Click **Tools/Cursor/SearchCommands...** , type *SFXVAL(1ms)*, and click **OK**. The cursor should position itself at 1ms and 1.6V. The cursor display box should show the cursor position. Click the right mouse button on the same marker and use the search command to find the voltage at 5ms. The results are shown in Figure 6-2. It should be apparent that the results obtained with PSpice for Windows agree favorably with the results obtained theoretically.

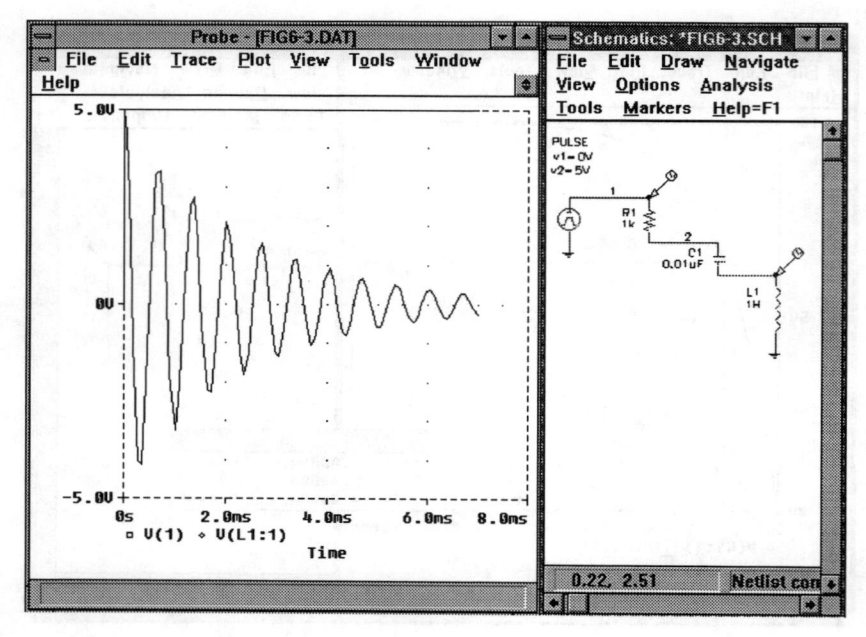

86   CHAPTER 6   RLC Circuits with Pulse Voltage Source (VPULSE)

**6-3  RLC circuit with VPULSE**

## 6.3  RLC CIRCUIT WITH VPULSE

Use *Schematics* to draw the circuit shown in Figure 6-3. Set a **transient analysis setup** as in section 6.2 (1ns print step, 7ms stop time) and Run PSpice. The applied voltage pulse and the voltage across the inductor are shown in Figure 6-3. Experiment with the circuit by changing the values of the components and observing the effects on the inductor voltage. PSpice for Windows allows you to do these experiments with an ease unequaled by any other method.

# Chapter 7  ELECTRONIC CIRCUITS

## 7.1 HALF-WAVE RECTIFIER

Figure 7–1 shows a half-wave rectifier. Use *Schematics* to draw the circuit. The transformer is **XFRM_LIN** from the **Analog** library. Double click the transformer symbol, double click **COUPLING=**, type *0.99*, click **Save Attr**, click **Change Display/Display Value/OK**. Double click **L1_VALUE=**, type *355mH*, and click **OK**. Double click **L2_VALUE=**, type *2.59mH*, and click **OK/OK**. The 15.3Ω resistor is the primary winding resistance. The 0.7Ω resistor is the secondary winding resistance. Set up a transient analysis with **PrintStep=** 0.1ms and **FinalTime=** 50ms.

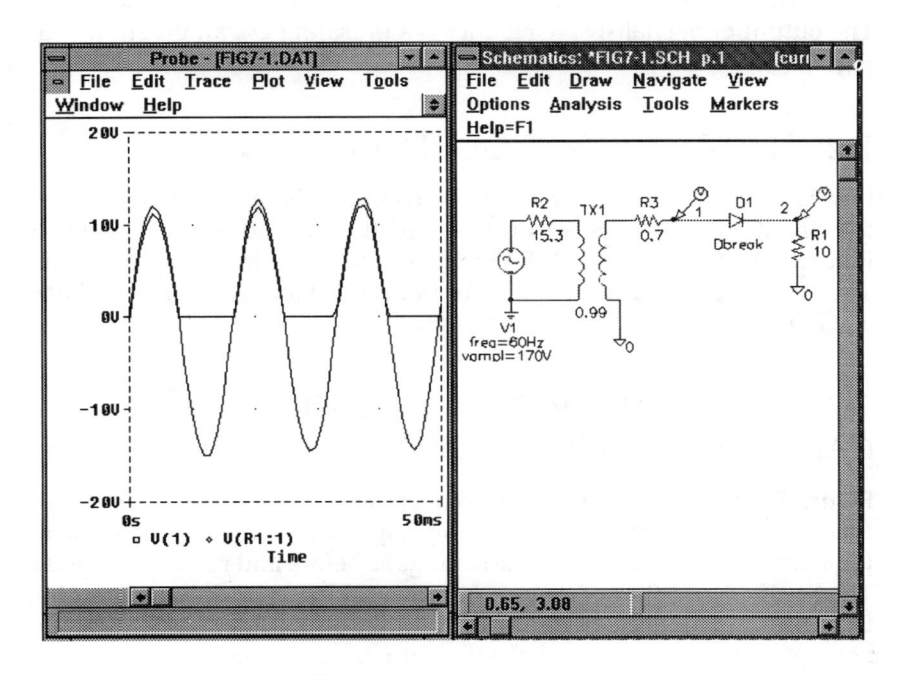

7–1  Half-wave rectifier

87

**7-2 Full-wave bridge rectifier**

Make sure you click the **Enable** check box for the transient analysis. To ensure that PROBE runs automatically after the analysis is completed, click **Analysis/Probe Setup/Auto Run Probe After Simulation/OK**. The output of the half-wave rectifier is a pulsating DC and is shown in Figure 7–1.

## 7.2 FULL-WAVE BRIDGE RECTIFIER

Figure 7–2 shows a full-wave bridge rectifier. Use *Schematics* to draw the circuit. Use the same settings as in section 7.1 and run the analysis. Figure 7–3 shows the PROBE display of the pulsating DC waveform. The + and – markers on the transformer secondary are the **Mark Voltage Differential** from the **Markers** menu.

## 7.3 FULL-WAVE BRIDGE RECTIFIER WITH CAPACITOR FILTER

Figure 7–4 shows a full-wave bridge rectifier with a capacitor filter. Use *Schematics* to modify Figure 7–2 by adding the 1000 microfarad capacitor across the load. Use the same settings as before and run the analysis. The rectified and filtered output is shown in Figure 7–5. Note that the DC voltage produced becomes constant at approximately 9.5V, with negligible ripple after a short transient period. Figure 7–6 shows the output with the capacitor changed to 100µF. Notice the increase in rip-

CHAPTER 7   Electronic Circuits

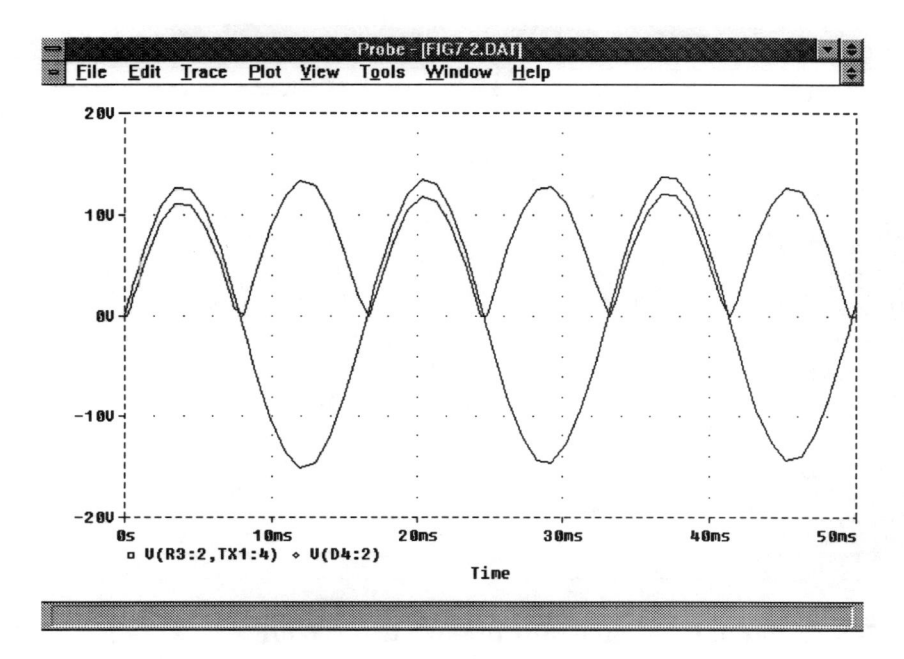

**7–3   PROBE display of full-wave rectifier output**

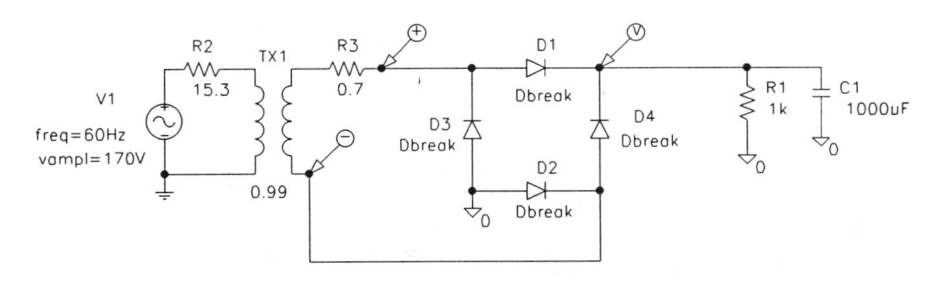

**7–4   Bridge rectifier with capacitor filter**

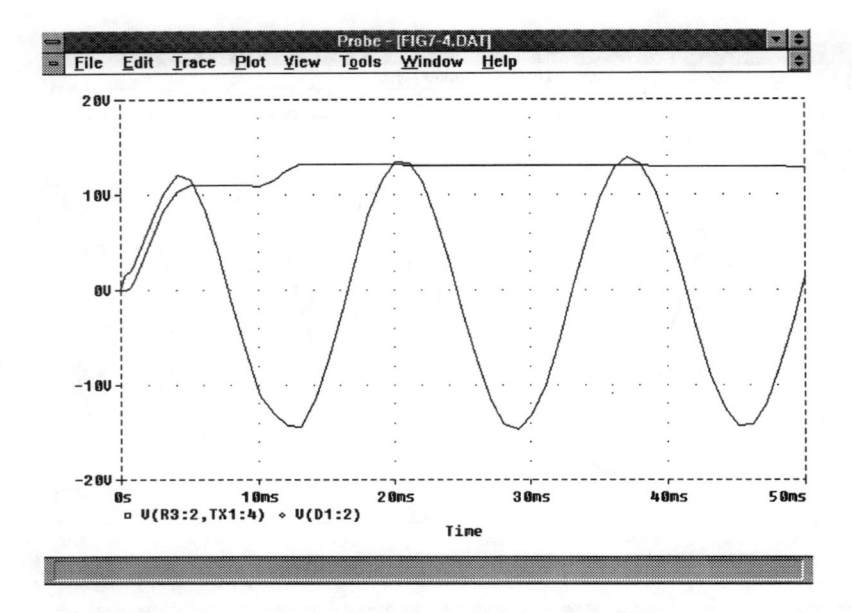

**7–5** Filtered full-wave output with C = 1000µF and RL = 1k

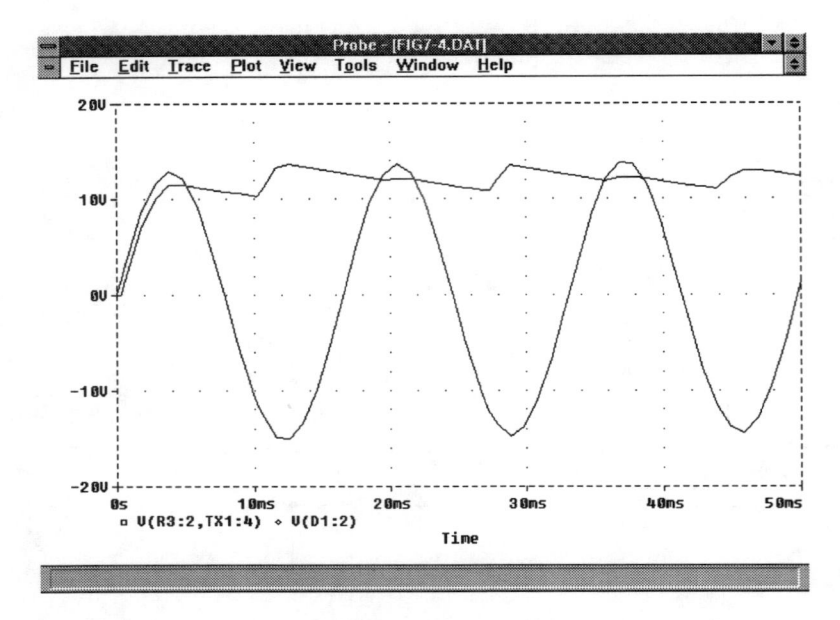

**7–6** Full-wave filtered output with C = 50µF and RL = 1k

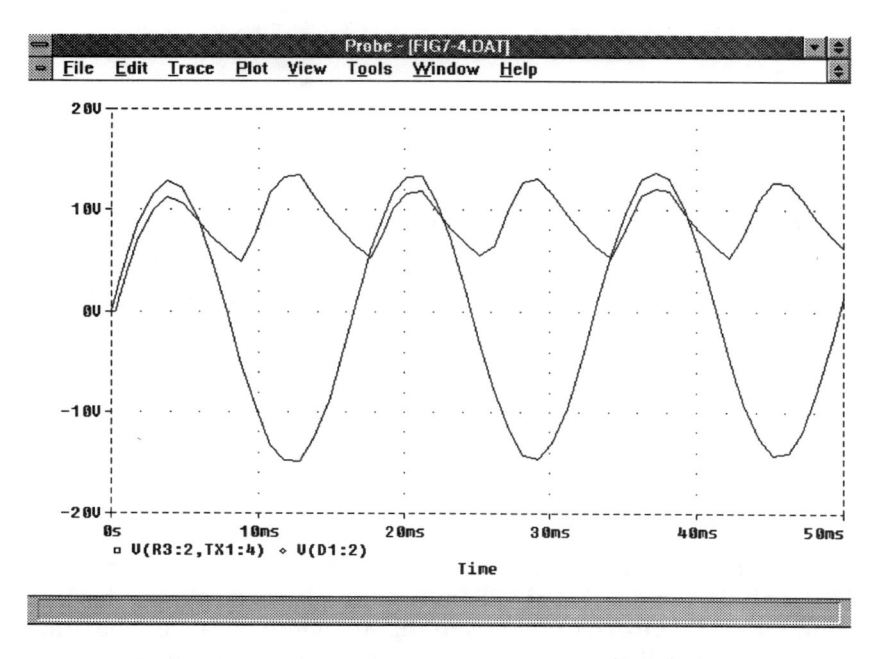

CHAPTER 7    Electronic Circuits    91

**7–7  Full-wave filtered output with C = 50µF and RL = 100Ω**

ple. Figure 7–7 shows the output with a small filter capacitor and a small load resistor. Notice the even larger ripple. Use *Schematics* to change the capacitor and resistor values and obtain the outputs shown here. Use the **FinalTime=** as indicated in the following figures.

## 7.4  BIPOLAR TRANSISTOR AMPLIFIER VOLTAGE DIVIDER BIAS

*Biasing* is the application of DC voltages to the terminal of electronic devices for the purpose of obtaining a desired operation. For a silicon NPN bipolar junction transistor to exibit transistor action ($I_C = \beta \cdot I_B$), the collector terminal must be at the highest voltage, the base terminal must be a few volts below the collector voltage, and the emitter terminal must be approximately 0.7V below the base voltage. A popular method of achieving these bias conditions is to use a battery and resistors. One circuit often used is called a *voltage divider bias circuit* and is shown in

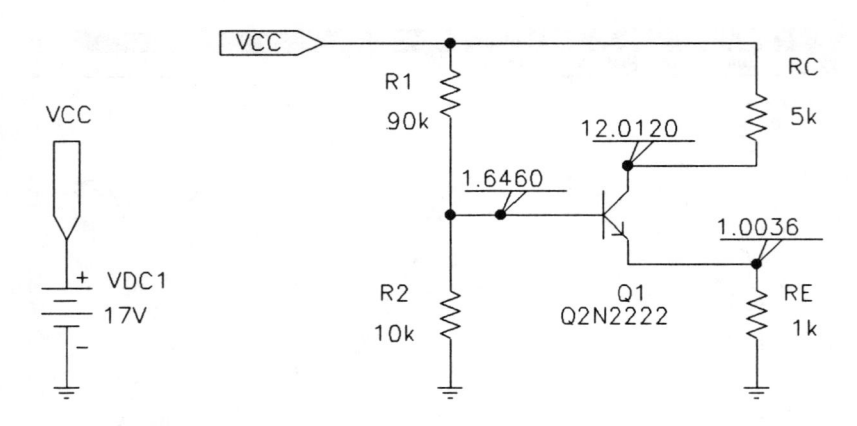

**7–8 BJT with voltage divider bias**

Figure 7–8. Use *Schematics* to draw the circuit and simply run PSpice. A DC analysis results. Figure 7–8 shows the bias conditions. The VCC connector is **GLOBAL** from the **Port** library. Once placed on the drawing, double click it and give it the label *VCC*. To connect any other node on the circuit to VCC, place a **GLOBAL** connector on the node and label it VCC. Note that the base voltage is 1.6V, which is approximately one-tenth of the supply voltage. The 10k, 90k voltage divider divides the supply voltage as follows:

$$VR1 = V_B = VCC \cdot \frac{R1}{(R1 + R2)} = 17 \cdot \frac{10k}{100k} = 1.7V$$

The emitter voltage is 0.7V below the base voltage because of the forward-biased silicon base-to-emitter junction, $V_E = V_B - 0.7 = 1.7 - 0.7 = 1V$. The emitter current is: $I_E = \frac{V_E}{RE} = \frac{1}{1k} = 1mA$. Because the collector current is almost equal to the emitter current, $I_c \approx 1mA$. This causes a 5 volt drop across the collector resistor, resulting in a 12 volt DC level at the collector. As can be seen in Figure 7–8, the PSpice results agree.

## 7.5 AN NPN BJT AMPLIFIER

Modify the circuit shown in Figure 7–8 by adding a signal source, a 10µF input coupling capacitor CC1, a 10µF output coupling capacitor

CHAPTER 7    Electronic Circuits                                              93

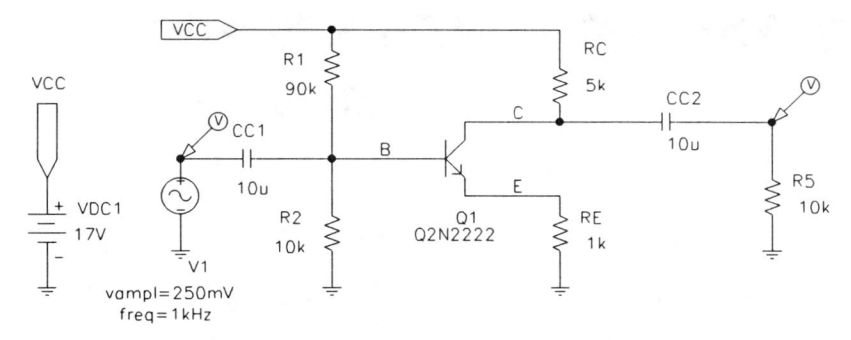

**7–9  BJT small signal amplifier**

CC2, and a 10k load resistor RL. The resulting circuit is shown in Figure 7–9. Use **VSIN** for the signal source and set it to an amplitude of 250mV and a frequency of 1kHz. Set up a transient analysis in 0.01ms steps and a 5ms final time. The PROBE display of the input signal and the output signal are shown in Figure 7–10. The signals and the associated DC levels may be displayed by moving the **Voltage Level** markers to the base and the collector terminals.

## 7.6  FREQUENCY RESPONSE OF A JFET AMPLIFIER

Figure 7–11 shows the circuit diagram of a JFET small signal amplifier. Use *Schematics* to draw the circuit. Set up the analysis as follows. Click **Analysis/Setup/AC Sweep/Decade**. Click inside the **TotalPts.:** text box, delete the contents, and type *10*. Click inside the **StartFreq.:** text box, delete the contents, and type *1Hz*. Click inside the **EndFreq.:** text box and type *1GHz*. Click **OK**. Activate the **enable** check box next to the **AC Sweep** button by clicking in the box at the left of the button and click **OK**. This sets up an analysis with 10 points per decade starting at 1 Hertz and ending at 1 gigaHertz.

Click **Run PSpice...** (name the circuit *Fig7–11*). When the analysis is finished, the PROBE window will open. Click **Trace/AddTrace** and type *db((absV(2))/(abs(V(1)))*. This command instructs PROBE to display the graph of the voltage gain from gate to drain in decibels. Click

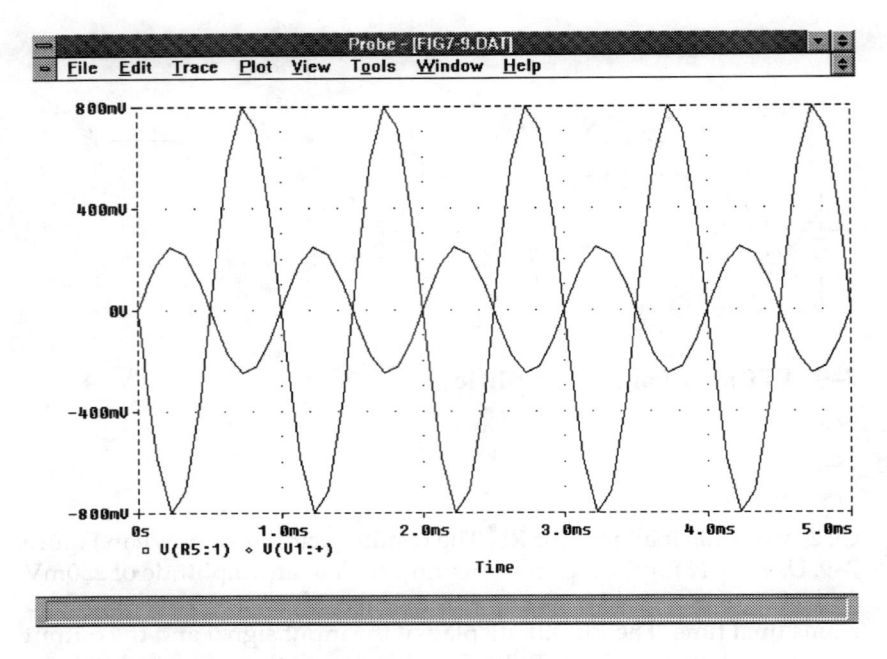

**7-10  PROBE display of BJT amplifier signals**

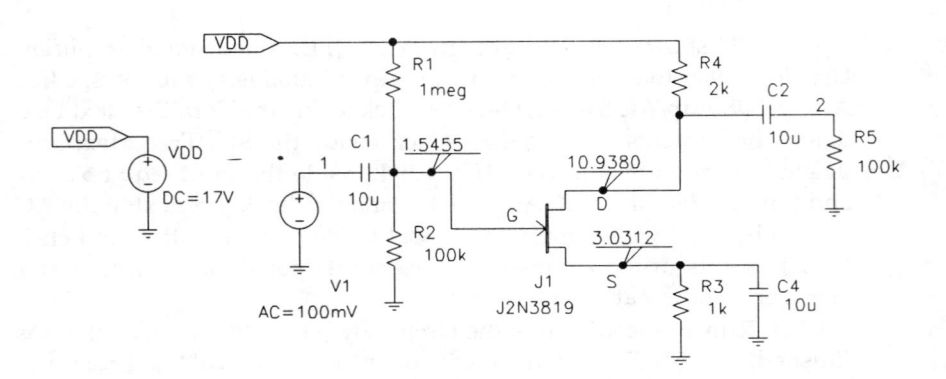

**7-11  JFET small signal amplifier**

CHAPTER 7   Electronic Circuits                                                      95

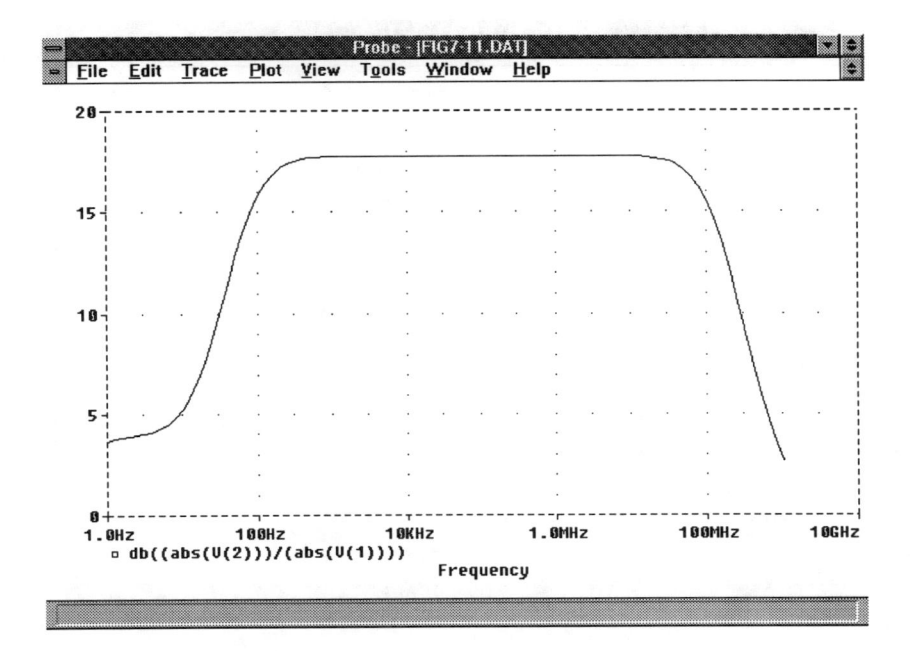

**7-12   PROBE display of the dB voltage gain from gate to drain**

**Plot/Add Plot/Trace/Add Trace** and type *P(V(2))*. This command instructs PROBE to add another plot and display the phase response. These two graphs are called the *frequency response*. The PROBE display is shown in Figure 7-12 and Figure 7-13.

## 7.7   FREQUENCY RESPONSE OF A TONE CONTROL

Figure 7-14 shows the circuit diagram for a tone control. At frequencies between 20Hz and 1kHz, the four capacitors show high impedance. With C3 basically open-circuited, the signal cannot get to the TREBBLE control pot. With C4 open, the TREBBLE control pot is disconnected from ground. The TREBBLE control pot has no effect on frequencies between 20Hz and 1kHz. C1 and C2 also show high impedance. The BASE control pot serves as a voltage divider

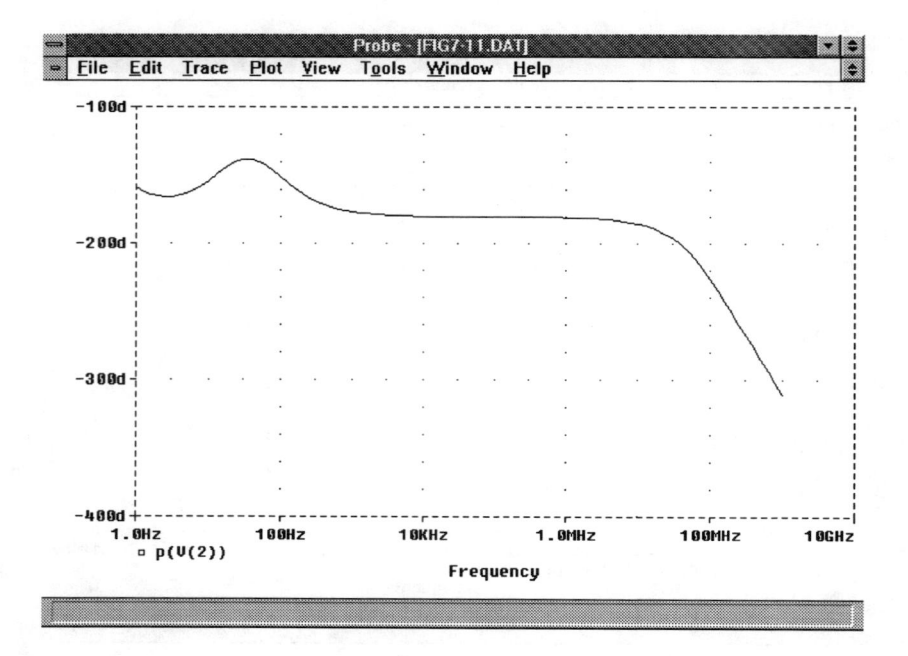

**7-13** PROBE display of the phase response

for frequencies between 20Hz and 1kHz. At frequencies between 1kHz and 20kHz, the four capacitors show low impedance. C1 and C2 basically provide a low impedance path around the BASE control pot. The BASE control pot has no effect on frequencies between 1kHz and 20kHz. The low impedance of C4 provides a ground for the bottom of the TREBBLE control pot, while the low impedance of C3 allows the signal to reach the top of the TREBBLE control pot. The TREBBLE control pot serves as a voltage divider for signals with frequencies between 1kHz and 20kHz. Figure 7-15 shows the PROBE display of the frequency response of the output as the BASE and TREBBLE controls are varied from maximum to minimum.

Use *Schematics* to draw the circuit for the tone control shown in Figure 7-14. Use **VSRC** (**source** library) for the signal source. Set **VSRC** at 1000mV. Use **POT** (**breakout** library) to draw the potentiometers. Set the value of the potentiometers to **50k**. The **SET=** parameter in the potentiometer attributes dialog box should be given the value *setval*. The

CHAPTER 7    Electronic Circuits    97

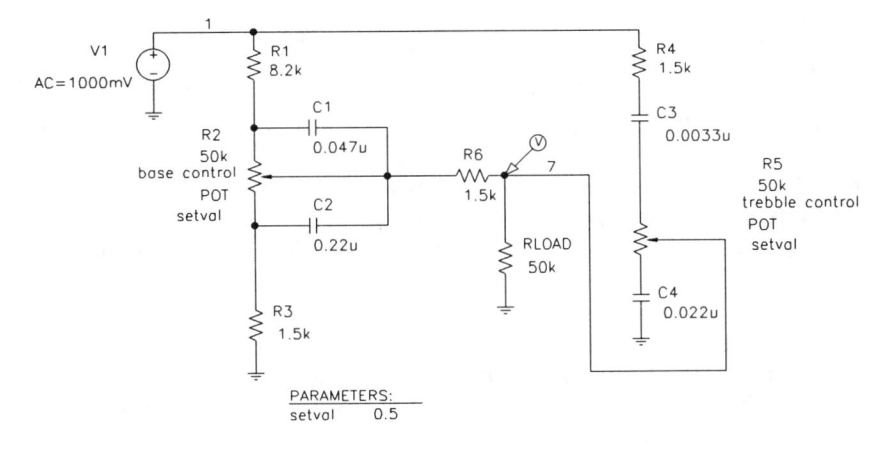

**7-14   Tone control**

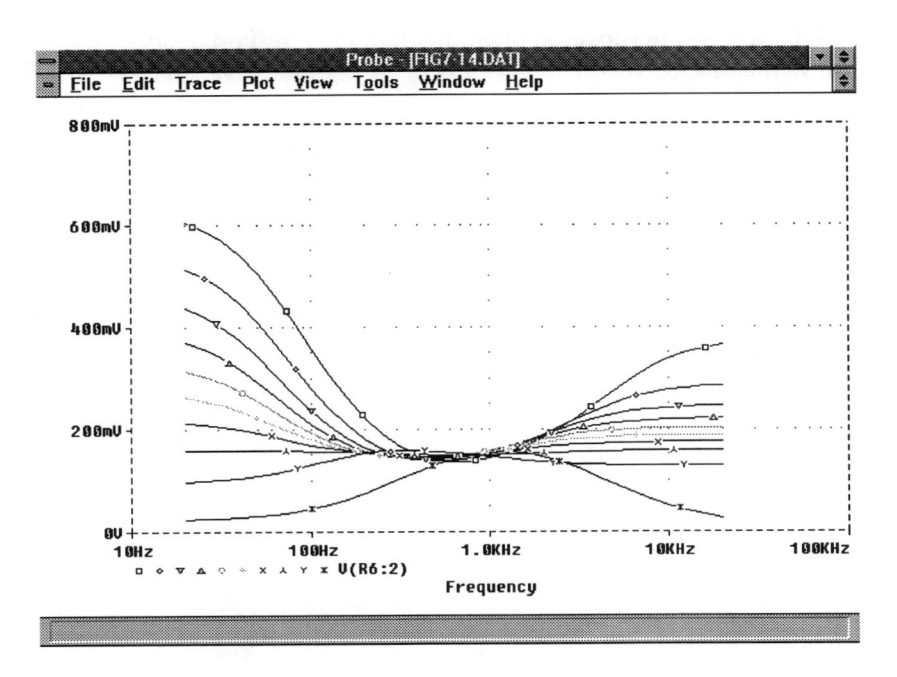

**7-15   PROBE display of tone control frequency response**

capacitors and resistors are drawn as usual. Change the capacitor and resistor designations and values as shown in Figure 7–14. When the circuit is wired, click **Draw/Get New Part/Special/PARAM/OK**. Double click the word **PARAMETERS:**. Double click **NAME1=**, type *setval*, and click **Save Attr**. Double click **VALUE1=**, type *0.5*, and click **Save Attr/OK**. Click **Analysis/Setup/AC Sweep/Decade**. Click inside the **Points/Decade** text box and change its contents to *10*. Click inside the **StartFreq** text box and change its contents to *20*. Click inside the **EndFreq** text box and change its contents to *20k*. Click **OK**. Enable the **AC Sweep** check box. This sets up a frequency sweep from 20 to 20kHz. Click **Parametric/Global Parameter/Linear**. Click inside the **Name** text box and type *setval*. Click inside the **StartValue** text box and type *0.1*. Click inside the **EndValue** text box and type *1*. Click inside the **Increment** text box and type *0.1*. Click **OK**. Enable the **Parametric** check box and click **OK**. This varies the value of the potentiometer parameter *setval* from 0.1 to 1 in 0.1 increments. This causes the potentiometers to vary from 5kΩ to 50kΩ in 5kΩ steps. PSpice performs an AC Sweep from 20Hz to 20kHz for each setting of the potentiometers. To run the analysis, click **Analysis/Run PSpice**. Name the circuit *Tone*. When the PROBE window opens, click **All/OK** in the **Available Sections** dialog box. PROBE will display a superposition of all the frequency responses. Figure 7–15 shows the PROBE display.

# Chapter 8  DIGITAL CIRCUITS

## 8.1 THE DIGITAL STIMULUS SOURCE STIM1 AND THE TTL INVERTER

*Schematics* offers a digital stimulus source called **STIM1**. Open *Schematics*, refer to Figure 8–1, and draw the circuit diagram as follows. Click **Draw/Get New Part/Browse/Source.slb/STIM1/OK**. Click to place the digital source STIM1 and click the right mouse button to exit the placement mode. Double click the **STIM1** symbol. Scroll down to COMMAND1. Double click **COMMAND1**. Type *0ms 0*. Click **Save Attr/Change Display/Value/OK**. Scroll to COMMAND2. Double click

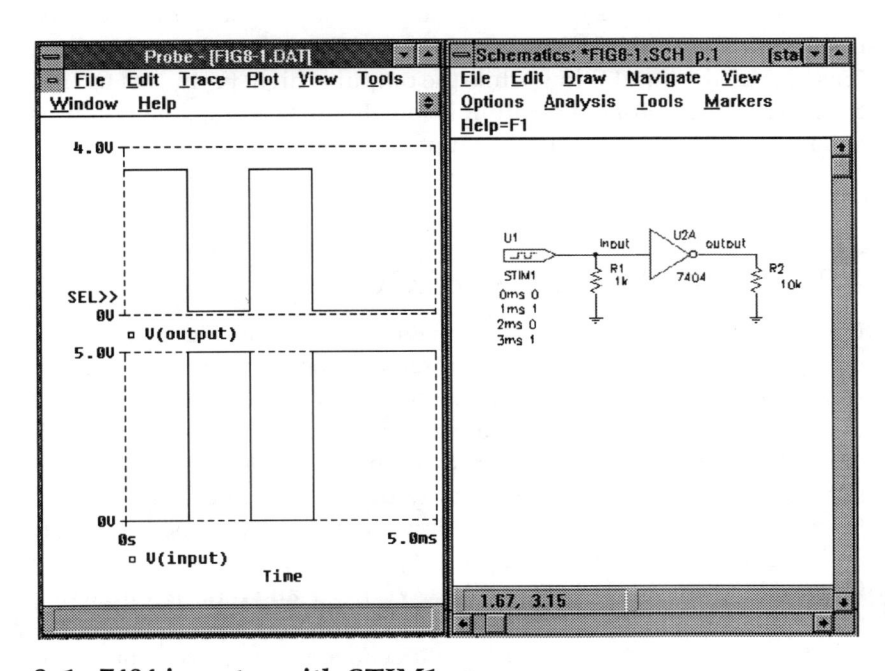

8–1  7404 inverter with STIM1 source

**COMMAND2**. Type *1ms 1*. Click **Save Attr/Change Display/Value/ OK**. Scroll to COMMAND3. Double click **COMMAND3**. Type *2ms 0*. Click **Save Attr/Change Display/Value/OK**. Scroll to COMMAND4. Double click **COMMAND4**. Type *3ms 1*. Click **SaveAttr/Change Display/Value/OK/OK**. Use click and drag to rearrange the drawing as shown in Figure 8-1. Continue drawing the circuit diagram by placing an inverter as follows. Click **Draw/Get New Part/Browse/7400.slb/ 7404/OK**. Place the inverter gate as shown. Place the two resistors as shown in Figure 8-1 and change the value of the load resistor to 10k. Place the ground symbols and draw the necessary wires to complete the circuit. Double click the wire at the input to the inverter and label it *input*. Double click the wire at the output of the inverter gate and label it *output*. To specify the analysis, click **Analysis/Setup.../Transient**. Change the contents of the **PrintStep** text box to *0.1ms* and the contents of the **FinalTime** text box to *5ms*. Click **OK**. Activate the **Enable** check box to the left of the **Transient** button. Click **OK/Analysis/Run PSpice**. Name the file *Fig8-1*. When the analysis ends, the PROBE window opens. If the PSpice window is open, minimize it. In the PROBE window, click **Trace/Add** and type *V(input)*. Click **Plot/Add Plot/Trace/ Add** and type *V(output)*. The input and the output wave forms will be displayed. The PROBE display is shown in Figure 8-1. Note that from 0ms to 1ms, the input is low while the output is high. From 1ms to 2ms, the input is high while the output is low. In other words, the output is always the logical inverse of the input.

## 8.2 TTL 7408 AND GATE LOGIC

Use *Schematics* to draw the logic diagram shown in Figure 8-2. Use drag and drop to position the different labels as shown. Use the same analysis setup as in section 8.1. The PROBE display of the inputs V(1) and V(2) and the output V(out) are shown in Figure 8-3. Note that the output is high only when both inputs are high, confirming the AND logic function. Any other logic gate function may be investigated using the same inputs as here. At 0ms, both V(1) and V(2) are 0 (making the input 0 0). At 1ms, V(1) is 0 while V(2) is 1 (making the input 0 1). At 2ms, V(1) is 1 and V(2) is 0 (making the input 1 0). Finally, at 3ms, both V(1) and V(2) are 1 (making the input 1 1). This is the standard input sequence for a two-input logic gate.

CHAPTER 8   Digital Circuits                                                101

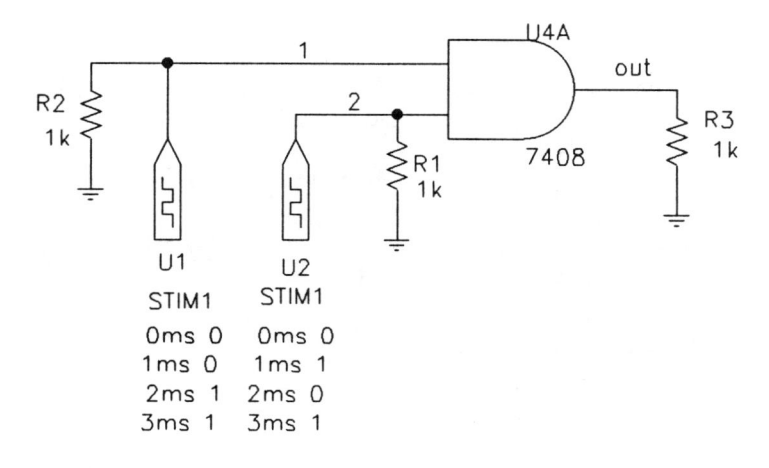

**8-2   7408 AND gate logic**

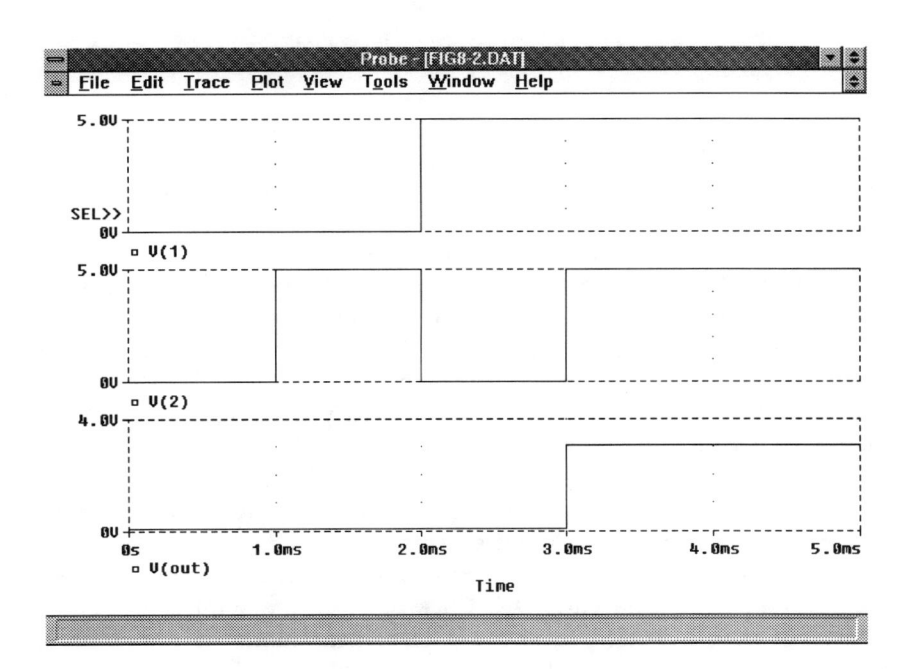

**8-3   PROBE display of TTL 7408 AND gate logic**

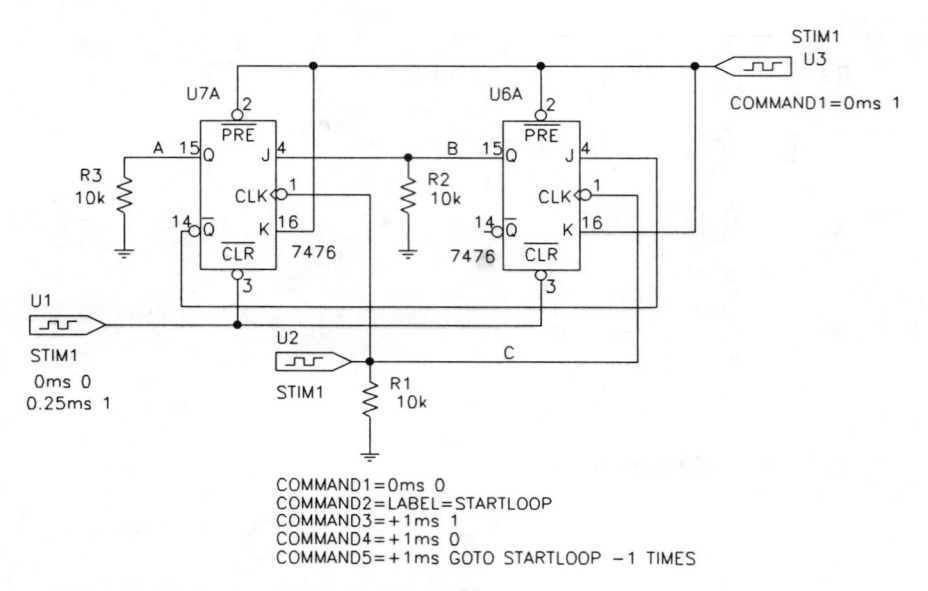

[Circuit diagram showing a modulo 3 synchronous counter using two 7476 JK flip-flops (U7A and U6A), with STIM1 sources U1, U2, U3, resistors R1, R2, R3 (10k each), and labeled nodes A, B, C.]

```
COMMAND1=0ms 0
COMMAND2=LABEL=STARTLOOP
COMMAND3=+1ms 1
COMMAND4=+1ms 0
COMMAND5=+1ms GOTO STARTLOOP -1 TIMES
```

**8-4 Modulo 3 synchronous counter**

## 8.3 MODULO 3 SYNCHRONOUS COUNTER

A counter that can assume three states is referred to as a "modulo 3" counter. A counter that updates all of its stages at the same time is referred to as synchronous. A synchronous modulo 3 counter can count 0, 1, 2, 0, 1, 2, 0, and so on. Figure 8-4 shows the circuit diagram for a synchronous modulo 3 counter. Use *Schematics* to draw the circuit diagram. **U1 STIM1** is a source providing a reset pulse at t = 0. Because the CLR is active low, the reset pulse source provides a low between 0ms and 0.25ms. At 0.25ms, the reset pulse source goes high and stays high. **U2 STIM1** is a digital source programmed to generate clock pulses 1ms wide with a 2ms period. To set up such a clock source, proceed as follows. Click **Draw/Get New Part**, type *STIM1*, and click **OK**. Click to place the symbol and click the right mouse button to end the placement mode. If the symbol is not labeled U2, change the label to U2. Double click the symbol. Scroll down to COMMAND 1. Double click **COMMAND1=** and type

CHAPTER 8    Digital Circuits    103

*0ms 0*. Click **Save Attr/Change Display/Value/OK**. Scroll to COMMAND2. Double click **COMMAND2=**. Type *LABEL=STARTLOOP*. Click **Save Attr/Change Display/Value/OK**. Scroll to COMMAND3. Double click **COMMAND3=**. Type *+1ms 1*. Click **Save Attr/Change Display/Value/OK**. Scroll to COMMAND4. Double click **COMMAND4=**. Type *+1ms 0*. Click **Save Attr/Change Display/Value/ OK**. Scroll to COMMAND5. Double click **COMMAND5=**. Type *+1ms GOTO STARTLOOP -1 TIMES*. Click **Save Attr/Change Display/Value/OK/OK**. Set up **U3 STIM1** to be high, starting at 0ms. Use two 7476 JK flip flops. Add the three 10k resistors and add wiring as needed. To complete the diagram, label wires A, B, and C as shown in Figure 8-4. A will be the most significant output bit, while B will be the least significant output bit. Wire C is the clock. Set up a transient analysis with a 1ms step and 20ms final time. Enable the transient analysis and run PSpice (name the file *Fig8-4*). When the PROBE window opens, click **Trace/Add** and type *V(B)*. Click **Plot/Add Plot/Trace/Add** and type *V(A)*. Click **Plot/Add Plot/Trace/Add** and type *V(C)*. The three PROBE traces show the

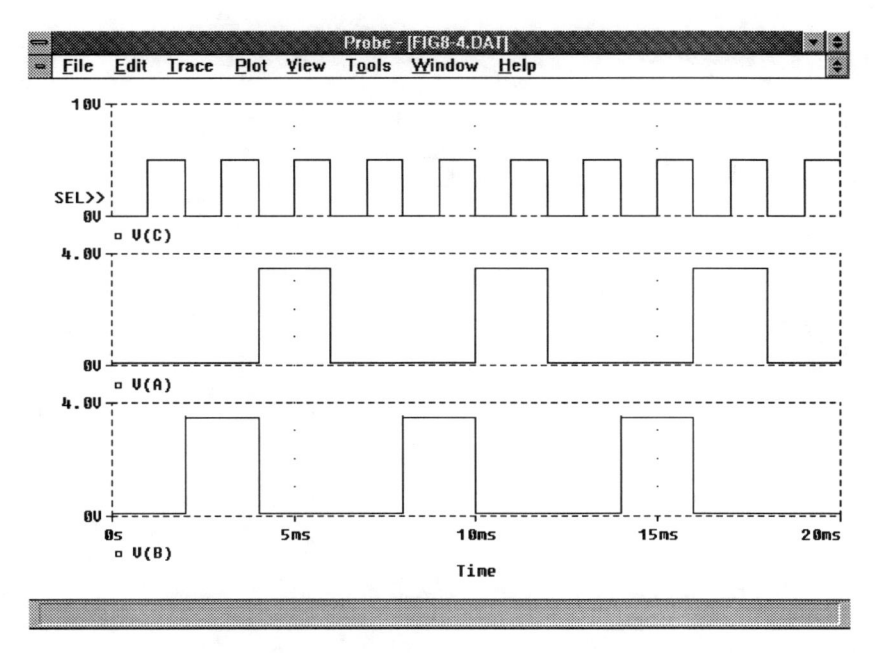

**8-5 PROBE display of timing diagram for modulo 3 synchronous counter**

clock and outputs A and B. This is referred to as a *timing diagram*. Note how the following truth table for the modulo 3 synchronous counter is implemented in the timing diagram (Figure 8-5).

| Clock | Output A | Output B |
|-------|----------|----------|
| 0 | low | low |
| 1 | low | high |
| 2 | high | low |
| 0 | low | low |

# Appendix A  IMPEDANCE CIRCUITS WITH AC SOURCES (PHASOR FORM)

## Chapter 4 for Version 5.3

(If your computer does not have a math co-processor and you are using Version 5.3 of PSpice for Windows use this appendix instead of the regular Chapter 4.)

## A.1  AC SOURCES

Electrical energy is usually generated and distributed by means of alternating currents and voltages. The general form of an alternating current waveform is:

$$i = I_{peak}\sin(\omega t \pm \varphi)\ amps \qquad \text{(Eq 4-1)}$$

Similarly for an alternating voltage:

$$v = V_{peak}\sin(\omega t \pm \varphi)\ volts \qquad \text{(Eq 4-2)}$$

Where:

$i$ = instantaneous current (A)
$I_{peak}$ = peak or maximum current (A)
$\omega$ = radian frequency (rad/s)
$\varphi$ = phase angle (degrees)
$v$ = instantaneous voltage (V)
$V_{peak}$ = peak or maximum voltage (V)

Although these waveforms are sinusoidal, it is customary to refer to them by their *phasor notation*, which consists of their EFFECTIVE value and PHASE ANGLE. The effective (or RMS) value for a sinusoid is: RMS = 0.707 • Peak. The equation for the typical house line voltage is:

$$v = 170\sin(377t + 0°) \text{ volts}$$

This voltage is usually referred to as $120\angle 0°V$. The 170V peak is 120V RMS and the phase angle is 0 because this waveform is considered reference. The frequency (60Hz or 377 rad/s) is not included in the phasor notation because the circuits considered do not alter the frequency. The phasor form of an electrical quantity is represented by placing a bar over its abbreviation. For our example:

$$\overline{V} = 120 \ 0° \text{ volts}$$

## A.2 PHASORS AND COMPLEX NUMBERS

Phasors are the polar form of complex numbers. Complex numbers may be represented in "polar form" (as a magnitude and an angle) or in "rectangular form" (as a real part and an imaginary part). The two forms are:

$$\overline{V} = c \angle \Theta \text{ volts} \qquad (Eq\ 4\text{-}3)$$

$$\overline{V} = (a + jb) \text{ volts} \qquad (Eq\ 4\text{-}4)$$

One form may be converted to the other by simple trigonometry and the Pythagorean theorem. If the polar form is known, then:

$$a = c \cdot \cos\Theta \qquad b = c \cdot \Theta$$

If the rectangular form is known then:

$$c = \sqrt{(a^2 + b^2)} \qquad \Theta = \tan^{-1}\left(\frac{b}{a}\right)$$

Complex number arithmetic applies to phasors. For addition and subtraction, it is simpler to use the rectangular form.

$$(a_1 + jb_1) + (a_2 + jb_2) = (a_1 + a_2) + j(b_1 + b_2)$$

$$(a_1 + jb_1) - (a_2 + jb_2) = (a_1 - a_2) + j(b_1 - b_2)$$

For multiplication and division, it is simpler to use the polar form.

$$(c_1 \angle \Theta_1) \cdot (c_2 \angle \Theta_2) = (c_1 \cdot c_2) \ (\Theta_1 + \Theta_2)$$

APPENDIX A   Impedance Circuits with AC Sources (Phasor Form)   107

$$\frac{(c_1 \angle \Theta_1)}{(c_2 \angle \Theta_2)} = (\frac{c_1}{c_2}) \angle (\Theta_1 - \Theta_2)$$

Most scientific calculators are capable of doing conversions fairly easily. Some of these calculators even handle complex number arithmetic (usually in rectangular form). Some calculators are capable of doing complex number arithmetic in mixed form.

In Chapter 5 we examine the response of circuits to sinusoidal inputs (time domain). In this chapter, we examine the response of circuits to inputs in phasor form (frequency domain).

In *Schematics*, VSRC can be used as a phasor voltage source and ISRC can be used as a phasor current source. VSRC can be used to measure AC currents.

## A.3   IMPEDANCE OF A RESISTOR

Impedance is represented by the letter Z with a bar over it ($\overline{Z}$). It represents the ratio of voltage to current in phasor form for a particular electrical device. Impedance has a magnitude and an angle and uses the units of ohms ($\Omega$). The impedance for a resistor is:

$$\overline{Z} = \frac{\overline{V_R}}{\overline{I_R}} = R \angle 0° \; \Omega$$

## A.4   RESISTOR WITH AN AC VOLTAGE SOURCE

The voltage across a resistor and the current through it are always in phase. The ratio of the voltage across the resistor to the current through it (impedance) is always the value of the resistance. A 10 volt source at 0 degrees, applied across a 2k$\Omega$ resistor, would cause a 5mA current at 0 degrees.

This can be shown in *Schematics* either by using the INCLUDE instruction or by examining the output in PROBE.

### A Resistor with an AC Voltage Source (using INCLUDE)

Draw the circuit diagram shown in Figure A-1 as follows. Click **Draw/Get New Part**. Type *VSRC* and click **OK**. Place the source as shown and click the right mouse button to end placement. Double click

# APPENDIX A  Impedance Circuits with AC Sources (Phasor Form)

**A-1  Circuit for AC current through resistor**

the source. Click **DC/Change Display/Display Value/Display Name/OK**. Double click **AC**. Type *10* <spacebar> 0 and click **Save Attr/Change Display/Display Value/Display Name/OK/OK**. Click **AC= 10 0** and click and drag to the right below **V1**. This shows that this is a 10 volt AC source at 0 degrees. Click **Draw/Get New Part**. Type *R* and click **OK**. Rotate the resistor 3 times with <Ctrl><R>. Place the resistor as shown in Figure A-1. Click the right mouse button. Click **R1**. Click **1k**. Click and drag **1k** below **R1**. Double click **1k**, type *2k*, and click **OK**. Click **Draw/Get New Part**. Type *VSRC* and click **OK**. Place the source V2 with the positive terminal touching the lower terminal of resistor R1. Click the right mouse button. Double click **V2**. Click **DC=/Change Display/Display Value/Display Name/OK/OK**. This source, with zero values for DC and AC, will serve as an AC ammeter. Click **Draw/Get New Part/Browse/Port/GND_EARTH/OK**. Place one ground symbol touching the lower terminal of V1 and another touching the right terminal of the ammeter labeled **V2**. Click the right mouse button. Click **Draw/Wire**. Click on the upper terminal of V1, stretch the wire to the upper terminal of R1, and double click. Click **Draw/Get New Part/ Browse/Special/INCLUDE/OK**. Place **INCLUDE** as shown in Figure A-1. Click the right mouse button. Click **File/Save**. Type *Fig4-1* and click **OK**. Click **Analysis/Examine Output**. This opens up NOTEPAD. Click **YES** to create a new file. To get a clear new file, click **File/New**. Type *.PRINT AC IM(V_V2) IP(V_V2)*. This statement instructs PSpice to show the magnitude (IM) and the phase (IP) of the current through the ammeter labeled V2. Click **File/Save** and type *C:\MSIMEV53\INC4-1.TXT*. Click **OK/File/Exit**. We have now generated the file which the **INCLUDE** statement will use in the analysis.

APPENDIX A   Impedance Circuits with AC Sources (Phasor Form)   109

Double click **INCLUDE**. Double click **FILENAME** and type *C:\MSIMEV53\INC4-1*.TXT. Click **Save Attr/OK**. To set up the analysis, click **Analysis/Setup../AC Sweep/Linear**. Click in the **Total Pts.:** box, delete the contents, and type *1*. Click in the **Start Freq.:** box, delete the contents, and type *1k*. Click in the **End Freq.:** box, delete the contents, and type *1k*. This sets up a frequency analysis consisting of 1 point starting at 1kHz and ending at 1kHz. Click **OK**. Click in the **Enabled** box to the left of the **AC Sweep** box to enable the AC Sweep. Click **OK**. To disable PROBE, click **Analysis/Probe Setup/Do Not AutoRun Probe/OK**. To run the analysis, click **Analysis/Run PSpice**. Type *Fig4–1* and click **OK**. When the analysis is finished, minimize the PSpice window. To see the results, click **Analysis/Examine Output**. Scroll down until the AC ANALYSIS section appears. Here the results are shown as: FREQ: 1.000E+03 IM(V_V2): 5.000E-03 IP(V_V2): 0.000E+00. These results indicate that at 1kHz, the current through the ammeter labeled V2 has a magnitude of 5mA and a phase of 0 degrees. A hard copy of the output file may be obtained by clicking **File/Print**. Click **File/Exit** to exit NOTEPAD.

## A Resistor with an AC Voltage Source (Using PROBE)

Basically the same results may be obtained by using PROBE. Use the preceding instructions to draw the circuit diagram shown in Figure A–2, but do not use the INCLUDE statement. To enable PROBE, click **Analysis/Probe Setup/Automatically Run Probe/OK**. To set up the analysis, click **Analysis/Setup/AC Sweep/Linear**. Click inside the **Total Pts.:** box, delete the contents, and type *2*. Click inside the **Start Freq.:** box,

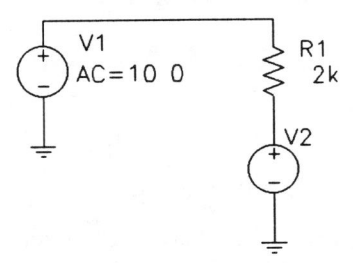

A–2   **Circuit for current through resistor using PROBE**

delete the contents, and type *1k*. Click inside the **End Freq.:** box, delete the contents, and type *2k*. Click **OK**. Click in the box to the left of the **AC Sweep** box to enable the AC Sweep. Click **OK**. Two points had to be selected because PROBE does not run with fewer than two data points. To run the analysis, click **Analysis/Run PSpice**. Type *FigA–2* and click **OK**. When the analysis is completed, the PROBE window opens. To obtain the magnitude data, click **Trace/Add**. Type *IM(V21)* and click **OK**. To obtain the phase data, click **Plot/Add Plot/Trace/Add**. Type *IP(V2)* and click **OK**. To use **Cursors** to pinpoint values click **Tools/Cursor/Display**. Click the left mouse button on the upper trace just to the right of the 1kHz mark. Tap <right cursor arrow> until the cursor B1 is at exactly 1kHz. The **Probe Cursor Window** shows that at 1kHz, the phase is 0.000 degrees. Click the right mouse button on the **IM(V2)** box of the lower trace. Click the right mouse button on the magnitude trace just to the right of the 1kHz mark. Hold <Shift> depressed while tapping <left cursor arrow> until cursor B2 is exactly on 1kHz. The **Probe Cursor Window** shows that B2 is at 1kHz and 5mA. To obtain a printout, click **File/Print/OK**.

## A.5 IMPEDANCE OF A CAPACITOR

The opposition of a capacitor to sinusoidally alternating current is referred to as *capacitive reactance*, is represented by the abbreviation $X_C$, and is measured in ohms. Capacitive reactance is inversely proportional to both the radian frequency and the capacitance value.

$$X_C = \frac{1}{\omega \cdot C}$$

In a capacitor, the current always leads the voltage by 90 degrees. The impedance of a capacitor is the ratio of the phasor voltage across the capacitor divided by the phasor current through the capacitor. It turns out that this ratio is also equal to the capacitive reactance at -90 degrees.

$$\overline{Z}_C = \frac{\overline{V}_C}{\overline{I}_C} = X_C \angle -90° \; \Omega$$

In rectangular form, capacitive impedance is:

$$\overline{Z}_C = (0 - jX_C) \; \Omega$$

A 0.1 microfarad capacitor at 1kHz would have a reactance of 1.592kΩ and an impedance of 1.592kΩ at -90 degrees in polar form (0-j1.592kΩ in rectangular form):

APPENDIX A    Impedance Circuits with AC Sources (Phasor Form)    111

$$X_C = \frac{1}{2 \cdot \pi \cdot f \cdot C} = \frac{1}{6280 \cdot 0.1\mu} = 1.592 k\Omega$$

$$\overline{Z}_C = 1.592 k\Omega \angle -90° = (0 - j1.592k)\Omega$$

If a 10 volt 0 degree source is applied across the capacitor, we would expect 6.281mA at 90 degrees.

$$\overline{I}_C = \frac{\overline{V}_C}{\overline{Z}_C} = \frac{10V\angle 0°}{1.592 k\Omega \angle -90°} = 6.281 mA \angle 90°$$

## A.6  CAPACITOR WITH AN AC VOLTAGE SOURCE (USING INCLUDE)

*Schematics* can be used to verify capacitive impedance. Use *Schematics* to draw the circuit diagram shown in Figure A-3. The procedure is similar to the one described for the resistor.

Once the analysis is completed, examine the output by means of **Analysis/Examine Output**. Scroll down to the AC ANALYSIS section and confirm that the current through the capacitor is 6.283mA at 90 degrees. Close NOTEPAD with **File/Exit**.

## A.7  IMPEDANCE OF AN INDUCTOR

The opposition of an inductor to sinusoidally alternating current is called *inductive reactance*, is represented by the abbreviation $X_L$, and is measured in ohms. Inductive reactance is directly proportional to both the radian frequency and the inductance value.

$$X_L = \omega L$$

**A-3  Circuit for AC current through a capacitor**

In an inductor, the current always lags the voltage by 90 degrees. The impedance of an inductor is the ratio of the phasor voltage across the inductor divided by the phasor current through the inductor. It turns out that this ratio is also equal to the inductive reactance at 90 degrees.

$$\overline{Z}_L = \frac{\overline{V}_L}{\overline{I}_L} = X_L \angle 90° = 0 + jX_L \ \Omega$$

A 0.5 Henry inductor at 1kHz has an impedance of 3.14kΩ at 90 degrees.

$$\overline{Z}_L = \omega L \angle 90° = (6280 \cdot 0.5) \angle -90° = 3.14k\Omega \angle 90°$$

If a 10 volt 0 degree voltage source is applied across this inductor, the resulting current would be 3.185mA at –90 degrees.

$$\overline{I}_L = \frac{10V \angle 0°}{3.14k\Omega \angle 90°} = 3.185mA \angle -90°$$

## A.8 INDUCTOR WITH AN AC VOLTAGE SOURCE (USING INCLUDE)

Use *Schematics* to draw and analyze the circuit diagram shown in Figure A–4. The procedure is similar to the one used earlier for the resistor. R1 = 0.0001Ω must be added in series with the inductor because PSpice does not accept a pure inductor across a voltage source.

After the analysis is concluded, examine the output file and confirm

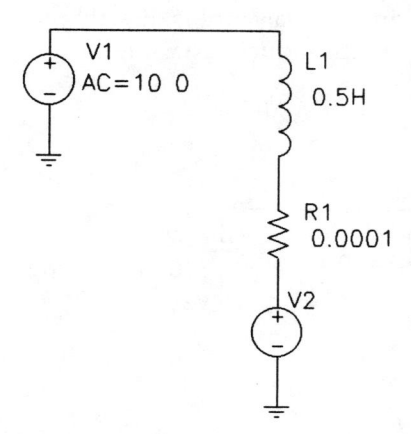

**A–4  Circuit for AC current through an inductor**

# APPENDIX A   Impedance Circuits with AC Sources (Phasor Form)    113

that the current through the inductor is 3.183mA at -90 degrees. Close NOTEPAD.

## A.9 IMPEDANCES IN SERIES WITH AC VOLTAGE SOURCE (USING INCLUDE)

Refer to Figure A–5. Impedances in series add. The impedance of the resistor is

$$\overline{Z}_R = (2k + j0) \; \Omega$$

while the impedance of the capacitor is

$$\overline{Z}_C = 1.592k\Omega \angle -90° = (0 - j1.592k)\Omega$$

The total impedance is the sum of the two:

$$\overline{Z}_T = \overline{Z}_R + \overline{Z}_C = (2k - j1.592k) \; \Omega = 2.56k\Omega \angle -38.52°$$

The current may be obtained by Ohm's Law:

$$\overline{I} = \frac{\overline{V}_1}{\overline{Z}_T} = \frac{10\angle 0°}{2.56k\angle -38.52°} = 3.91mA \angle 38.52°$$

The voltage across R1 may be obtained by Ohm's Law:

$$\overline{V}_{R1} = \overline{I} \cdot \overline{Z}_R = (3.91mA\angle 38.52°)(2k\angle 0°) = 7.82V\angle 38.52° = (6.12 + j4.87)V$$

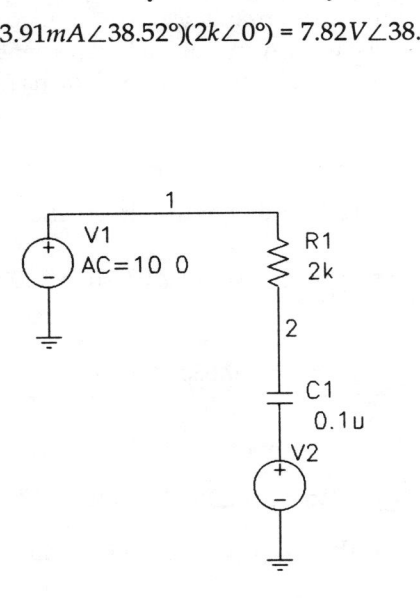

**A–5   Impedances connected in series**

The voltage across the capacitor may be obtained by using KVL:

$\overline{V}_C = \overline{V}_1 - \overline{V}_{R1} = (10+j0) - (6.12 + j4.87) = (3.88 - 4.87)V = 6.23V\angle -51.48°$

These results may be confirmed with *Schematics*. Use *Schematics* to draw and analyze the circuit shown in Figure A–5. Use the procedure outlined in section A.4 and use NOTEPAD to write and save the file C:\msimev53\inc4-5.txt. This file should consist of the following line:

**.PRINT AC IM(V_V2) IP(V_V2) VM(2) VP(2)**

Use this file with the INCLUDE instruction as shown in Figure A–5. Use the same AC Sweep as in section A.4. After the analysis is concluded, examine the output file. The output file should confirm that the magnitude of the current through the ammeter labeled V2, IM(V_V2), is 3.91mA and its phase, IP(V_V2), is 38.52 degrees. The magnitude of the voltage at node 2, VM(2), is 6.23V while its phase, VP(2), is -51.48 degrees. Note that the voltage at node 2 is also the voltage across the capacitor.

## A.10 IMPEDANCES IN PARALLEL WITH AC VOLTAGE SOURCE (USING INCLUDE)

Admittance is the reciprocal of impedance. Admittance is to impedance as conductance is to resistance. Admittance is represented by an upper case Y with a bar over it ($\overline{Y}$) and is measured in Siemens (abbreviated: S).

$$\overline{Y} = \frac{1}{\overline{Z}}$$

Admittances in parallel add. Refer to the circuit shown in Figure A–6.

$$\overline{Y}_{R1} = \frac{1}{\overline{Z}_{R1}} = \frac{1}{2k\Omega\angle 0°} = 0.5mS\angle 0° = (0.5 + j0)mS$$

$$\overline{Y}_{C1} = \frac{1}{\overline{Z}_{C1}} = \frac{1}{1.592k\Omega\angle -90°} = 0.628mS\angle 90° = (0 + j0.628)mS$$

The total admittance is the sum of the two.

$$\overline{Y}_T = \overline{Y}_{R1} + \overline{Y}_{C1} = (0.5 + j0.628)mS = 0.803mS\angle 51.474°$$

The total impedance is the reciprocal of the total admittance.

$$\overline{Z}_T = \frac{1}{\overline{Y}_T} = \frac{1}{0.803mS\angle 51.474°} = 1.245k\Omega\angle -51.474°$$

APPENDIX A   Impedance Circuits with AC Sources (Phasor Form)   115

Another way of obtaining the total impedance is to use the "product over sum" formula:

$$\overline{Z}_T = \frac{\overline{Z}_{R1} \cdot \overline{Z}_{C1}}{\overline{Z}_{R1} + \overline{Z}_{C1}} = \frac{(2k\Omega\angle 0°)(1.592k\Omega\angle -90°)}{(2 - j1.592)k\Omega} = 1.245k\Omega\angle -51.474°$$

The total current may be obtained by Ohm's Law:

$$\overline{I}_T = \frac{\overline{V}}{\overline{Z}_T} = \frac{10V\angle 0°}{1.245k\Omega\angle -51.474°} = 8.032mA\angle 51.474° = (5.003 + j6.284)mA$$

The current through each component may also be found individually by Ohm's Law:

$$\overline{I}_{R1} = \frac{\overline{V}_1}{\overline{Z}_{R1}} = \frac{10V\angle 0°}{2k\Omega\angle 0°} = 5mA\angle 0° = (5 + j0)mA$$

$$\overline{I}_{C1} = \frac{\overline{V}_1}{\overline{Z}_{C1}} = \frac{10V\angle 0°}{1.592k\Omega\angle -90°} = 6.281mA\angle 90° = (0 + j6.281)mA$$

The total current may be obtained by KCL:

$$\overline{I}_T = \overline{I}_{R1} + \overline{I}_{C1} = (5 + j6.281)mA = 8.028mA\ 51.478°$$

*Schematics* may be used to confirm these results. Use *Schematics* to draw and analyze the circuit diagram shown in Figure A–6. Generate the file C:\MSIMEV53\INC4-6.TXT as described previously. This file should contain one line:

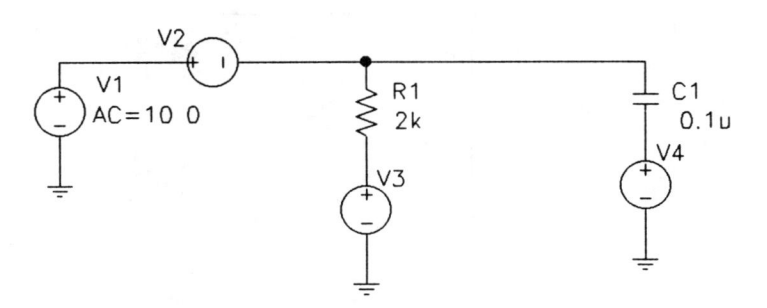

.PRINT AC IM(V_V2) IM(V_V2) IM(V_V3) IP(V_V3) IM(V_V4) IP(V_V4)

A–6  Impedances connected in parallel

116    APPENDIX A    Impedance Circuits with AC Sources (Phasor Form)

Use this file in the INCLUDE instruction as shown in Figure A–6. Set up the AC Sweep and run the analysis exactly as in the previous sections.

At the completion of the analysis, examine the output file. The three currents obtained earlier should check favorably with the currents obtained with *Schematics*.

## A.11 IMPEDANCES IN SERIES-PARALLEL WITH AC VOLTAGE SOURCE (USING INCLUDE)

Refer to the circuit diagram shown in Figure A–7. The impedance of R2 is negligible and is included here because PSpice does not accept "pure" inductors.

$$\overline{Z}_{R1} = 2k\Omega \angle 0° \qquad \overline{Z}_{C1} = 1.592k\Omega \angle -90° \qquad \overline{Z}_{L1} = 3.14k\Omega \angle 90°$$

$$\overline{Z}_X = \overline{Z}_{C1} // \overline{Z}_{L1} = \frac{(1.592k\Omega \angle 90°)(3.14k\Omega \angle 90°)}{(-j1.592k\Omega) + (j3.14k\Omega)}$$

$$= 3.229k\Omega \angle -90° = -j3.229k\Omega$$

$$\overline{Z}_T = \overline{Z}_{R1} + \overline{Z}_X = (2 - j3.229)k\Omega = 3.798k\Omega \angle -58.228°$$

The total impedance may be obtained by using *Schematics* and "Brumgnach's 1 amp rule." Figure A–7 shows the circuit with a 1A at 0

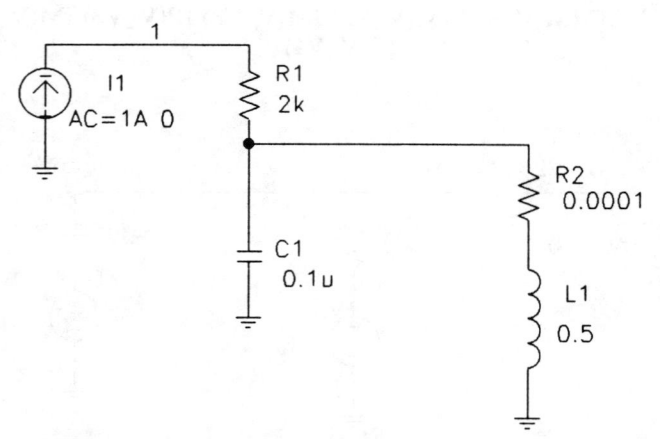

**A–7  Total impedance obtained with *Schematics***

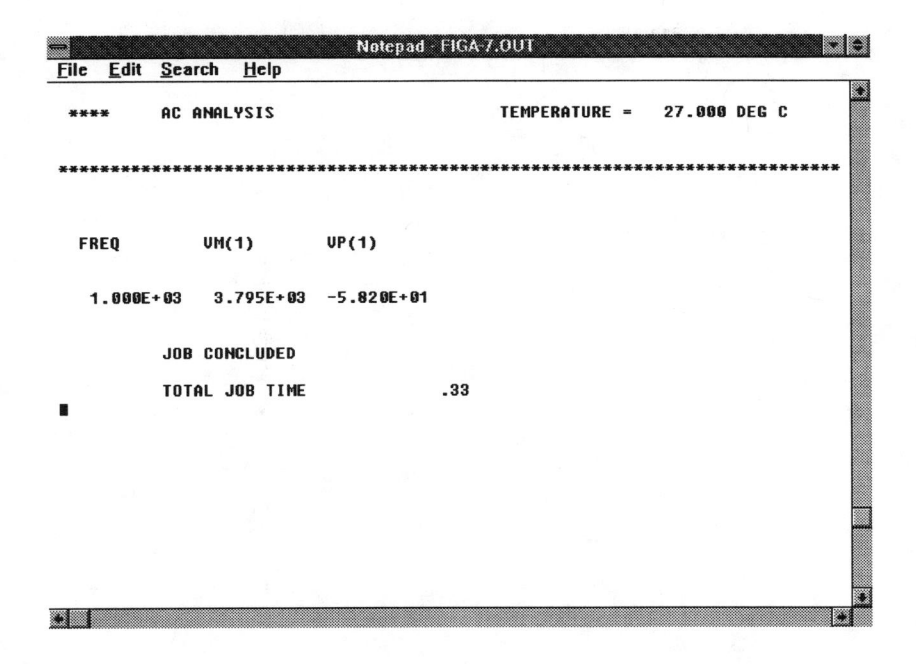

APPENDIX A   Impedance Circuits with AC Sources (Phasor Form)   117

```
                        Notepad - FIGA-7.OUT
File  Edit  Search  Help

****       AC ANALYSIS                    TEMPERATURE =   27.000 DEG C

*****************************************************************


    FREQ          VM(1)        VP(1)

    1.000E+03    3.795E+03    -5.820E+01

              JOB CONCLUDED
              TOTAL JOB TIME              .33
```

**A–8   Section of output file showing value of** VM(1) **and** VP(1)

degrees current source. The **INCLUDE** instruction to be added to the diagram should consist of the following line:

**.PRINT AC VM(1) VP(1)**

The instruction is set up as the file C:\MSIMEV53\INC4-7.TXT, as described previously in section A.4. This instruction tells the program to print the magnitude and phase of the voltage at node 1. Because the impedance is equal to the voltage divided by the current and the current is 1A, the total impedance in ohms is the value of the voltage at node 1. (Division by 1 does not alter the value of the numerator.) The value of the **VM(1)** and **VP(1)** may be seen by clicking **Analysis/Examine Output** and scrolling down to the AC ANALYSIS section. Figure A–8 shows the section of the output file containing the value of Ztotal.

The total current may be found by Ohm's Law.

$$\bar{I}_T = \frac{\bar{V}_1}{\bar{Z}_T} = \frac{10V\angle 0°}{3.798k\Omega\angle -58.228°} = 2.633mA\angle 58.228° = (1.386 + j2.238)mA$$

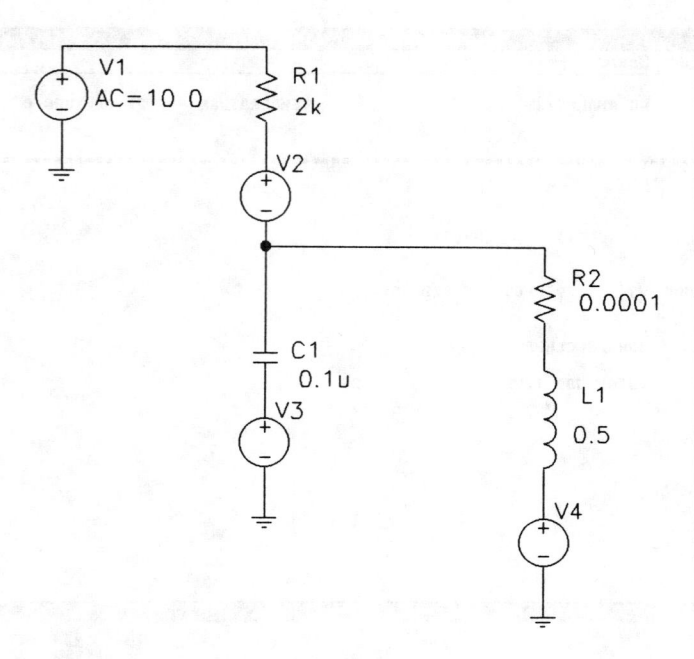

## APPENDIX A  Impedance Circuits with AC Sources (Phasor Form)

[Circuit diagram: V1 AC=10 0; R1 2k; V2; C1 0.1u with V3; R2 0.0001; L1 0.5; V4]

### A-9  Impedances in series-parallel

The current through the capacitor and the inductor may be found by current division.

$$\bar{I}_{C1} = \bar{I}_T \frac{\bar{Z}_{L1}}{\bar{Z}_{L1} + \bar{Z}_{C1}} = (2.633 mA \angle 58.228°) \frac{3.14 k\Omega \angle 90°}{(-j1.592 k\Omega) + (j3.14 k\Omega)}$$

$$= (2.812 + j4.54) mA = 5.34 mA\ 58.228°$$

$$\bar{I}_{L1} = \bar{I}_T \frac{\bar{Z}_{C1}}{\bar{Z}_{C1} + \bar{Z}_{L1}} = (2.633 mA \angle 58.228°) \frac{1.592 k\Omega \angle -90°}{(-j1.592 k\Omega) + (j3.14 k\Omega)}$$

$$= (-1.426 - j2.303) mA = 2.708 mA \angle -121.772°$$

The total current may also be found by KCL.

$$\bar{I}_T = \bar{I}_{C1} + \bar{I}_{L1} = (2.812 + j4.54) mA + (-1.426 - j2.302) mA$$

$$= (1.386 + j2.238) mA = 2.632 mA \angle 58.23°$$

Use *Schematics* to draw and analyze the circuit shown in Figure A-9.

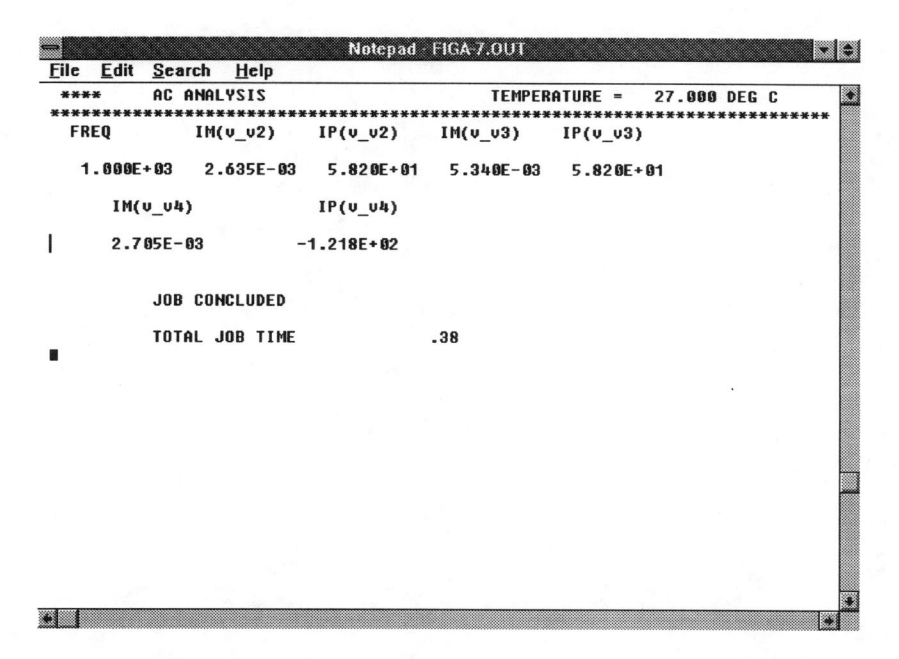

# APPENDIX A    Impedance Circuits with AC Sources (Phasor Form)    119

```
              Notepad - FIGA-7.OUT
File  Edit  Search  Help
****       AC ANALYSIS                    TEMPERATURE =   27.000 DEG C
*************************************************************************
    FREQ        IM(v_v2)      IP(v_v2)     IM(v_v3)     IP(v_v3)

   1.000E+03    2.635E-03    5.820E+01    5.340E-03    5.820E+01

         IM(v_v4)            IP(v_v4)
|        2.705E-03           -1.218E+02

            JOB CONCLUDED
            TOTAL JOB TIME              .38
■
```

**A-10**  AC portion of output file for circuit shown in Figure A-9

The **INCLUDE** instruction should be the file made up of the following line:

**.PRINT AC IM(V_V2) IP(V_V2) IM(V_V3) IP(V_V3) IM(V_V4) IP(V_V4)**

This should be set up as the file *C:\MSIMEV53\INC4-9.TXT*, as described previously. Use the same AC Sweep as before. When the analysis is completed, examine the AC ANALYSIS portions of the output file and confirm that the three currents obtained with PSpice agree with the ones obtained earlier. The results are shown in Figure A-10.

# Appendix B  MAKING AN OHMMETER (RMeter, RProbe)

# MAKING AN IMPEDANCE METER (ZMeter, ZProbe)

## B.1 MAKING AN OHMMETER (RMeter)

In PSpice for Windows DC voltages can be measured with a **VIEWPOINT** while DC currents can be measured with an **IPROBE**. Both of these devices are found in the **special** library. There is however no symbol that measures resistance directly. The purpose of this section is to design an ohmmeter which will be named **RMeter**. RMeter will be stored in the **Brum** library. If you did not set up the **Brum.slb** library in Chapter 1 or if you prefer, you can store the new devices in the **special.slb** library. To do this, substitute **special.slb** every place you see **Brum.slb** and if you are ever asked "Save to present library?" click **YES**.

To open the library editor and the special library, click **File/Edit Library/File/Open**. Scroll down until **special.slb** is visible, and click **special.slb/OK**. We will now copy the VIEWPOINT symbol into the library editor. Click **Part/Copy**. Scroll down until **VIEWPOINT** is visible. Click **VIEWPOINT**. Next, we will assign the name RMeter to the ohmmeter symbol that well make shortly. Click inside the **New Part Name** box and delete the contents. In the (now blank) **New Part Name** box type *RMeter*. Click **OK**. We will now delete the VIEWPOINT symbol. Select the horizontal line of the symbol by clicking on it and tap the <Del> key

122    APPENDIX B    Making an Ohmmeter/Impedance Meter

to delete it. Select the other two lines of the symbol and delete them. Select the remaining pin (denoted by an x) and delete it.

We will now do some graphics. To ensure that all lines are on the grid, click **Options/Display Options**. Click on any option that has an empty check box—every box should have an X in it. Click **OK**. The bounding box will now be enlarged. Click **Graphics/Bbox**. Move the pencil to a point 8 dots to the right and 6 dots down from the upper left hand corner of the bounding box. Double click at that point. We will now draw a box for our new ohmmeter. Click **Graphics/Box**. Click the pencil at a point 1 dot below the upper left hand corner of the bounding box. Move the pencil to a point 1 dot above the lower right hand corner and click. We will now place two connection pins. Click **Graphics/Pin**. Rotate the pin symbol (a line and an x) three times by tapping <Ctrl><R> three times. Move the pin symbol to a spot 4 dots to the right of the upper left hand corner of the bounding box. Click to place the pin. Rotate the pin symbol two times. Move the pin symbol to a spot 4 dots to the left of the lower right hand corner of the bounding box. Click to place the pin. Click the right mouse button to end pin placement.

We will now remove the labels from the pins to clear up the symbol. Double click the wire of pin 1. Click **Display Name**. The box should now be empty. Click **OK**. Repeat for pin 2. We will now move the numbers 1 and 2 to more appropriate locations. Click the number 1 to select it. Drag the number 1 to the right of the upper pin. Select the number 2 and drag it to the right of the lower pin. Next, we will make pin 2 the reference pin. Click **Graphics/Origin**. Move the pencil to the x on pin 2 and double click.

Next, we will add two attributes. Click **Part/Attributes**. Click inside the **Name** box and type *REFDES*. Click inside the **Value** box and type *RMeter*. Click **Display Value/Save Attr**. If a dialog box opens up saying: "This attribute normally goes on a predefined layer. Override?" click **yes**. We will next put a 1 amp current source inside our symbol forcing current out of the symbol at pin 1. Click inside the **Name** box and delete the contents. Type *TEMPLATE*. Click inside the **Value** box and delete the contents. Type *I^@REFDES<spacebar>%pin2<spacebar>%pin1<spacebar>1* making sure to leave spaces only where <spacebar> is indicated. If the **Display Value** checkbox has an X in it, click on the box to clear the X. Click **Save Attr/OK**. We will now move the name RMeter inside the ohmmeter box. Select the word RMeter by clicking on it. Drag it to the upper middle of the box. We will next insure that the resistance readings are displayed in the box, under the word RMeter. Click **Part/Attrib/BIASVALUE=/Display Name/Save Attr/OK**.    Click

APPENDIX B   Making an Ohmmeter/Impedance Meter        123

BIASVALUE= to select it and drag it to a location inside the box under the word RMeter. We can now hide the word BIASVALUE= so that only the resistance value will show at that location but not the word BIASVALUE=. Click **Part/Attributes/BIAS VALUE=/Display Name**. The box should now be empty. Click **Save Attr/OK**.

To save the new symbol, click **Part/Save Changes/Part/Save to Library/Brum.slb/OK**. Click **File/Return to Schematics**. If asked "Save to current library?" click **no**. The ohmmeter named RMeter is now available in the Brum.slb library. For correct readings, pin 2 of RMeter must always be grounded.

## B.2   USING RMeter

We will now use RMeter to measure the total resistance of a circuit. You must keep in mind that to measure the total resistance of a circuit, all voltage sources must be replaced by short circuits and all current sources must be replaced by open circuits. In Chapter 3, Figure 3-5 shows Brumgnach's 1 amp rule used to find the total resistance of the circuit. We will now use RMeter instead.

In Figure B-1, R2 is in parallel with R3 giving an equivalent resistance of 2k2. The parallel combination is then in series with R1 for a total resistance of 7k2. Use schematics to draw the circuit shown in Figure B-1. Run PSpice to analyze the circuit. Figure B-2 shows RMeter indicating the total resistance.

## B.3   MAKING RMeter INTO RProbe

The symbol RMeter requires two connections when placed in a circuit, one connection to the circuit to be measured and one connection to ground. RMeter can be modified so that it needs only the connection to the circuit it is measuring. The ground connection may be included in

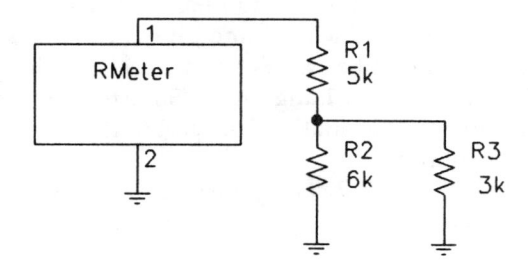

**B-1   Circuit with WMeter**

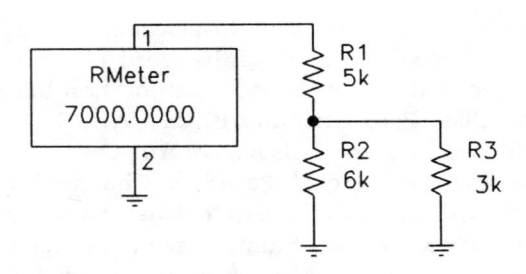

**B-2 RMeter showing total resistance**

the symbol and the grounded pin can be made invisible. The new symbol will show only one pin and will require only one connection to the circuit. The one pin symbol will be named **RProbe**. RProbe will always give the total resistance from its connection point to ground.

Click **File/Edit Library/File/Open**. Scroll down until **brum.slb** is visible (assuming you saved RMeter in the brum.slb library). Click **brum.slb/OK**. Click **Part/Copy/RMeter**. Click inside the **New Part Name** box and delete its contents. Type *RProbe*. Click **OK**. Click **Part/Attributes/REFDES**. Click inside the **Value** box and delete its contents. Type *RProbe*. Click **Save Attr/OK**. We will now exchange the positions of pin 1 and pin 2. Click on pin 2 and drag it to the right outside the drawing. Rotate pin 2 twice (tap <Ctrl><R> twice). Click pin 1 and drag it to the right outside the drawing. Rotate pin 1 twice. Drag pin 1 to the old location of pin 2. Click pin 2. Drag pin 2 to the old location of pin 1. Click **Graphics/Origin** and double click the pencil on the x of pin 2. We can now hide pin 2 and connect it to ground. Double click on pin 2. Click in the **Hidden** check box (the box should now have an X in it). Click inside the **Net** box and type the number 0. Click **OK**. Double click pin 1. Click **Edit Attributes**. Click the line **pin=1**. Click **Display Value** to clear the X so the pin number won't show in schematics. Click **Save Attr/OK/OK**. Click **Part/Save Changes/Part/Save to Library/brum.slb/OK**. Click **File/Return to Schematic**. The symbol **RProbe** is now available in the brum.slb library.

Figure B-3 shows RProbe indicating the total resistance of a resistive circuit.

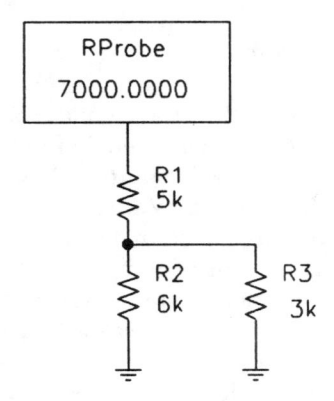

APPENDIX B    Making an Ohmmeter/Impedance Meter    125

```
        ┌──────────────┐
        │   RProbe     │
        │  7000.0000   │
        └──────┬───────┘
               │
              ┌┴┐
              │ │ R1
              │ │ 5k
              └┬┘
               ├──────────┐
              ┌┴┐        ┌┴┐
              │ │ R2     │ │ R3
              │ │ 6k     │ │ 3k
              └┬┘        └┬┘
               ⏚          ⏚
```

**B-3   RProbe showing total resistance**


## B.4   MAKING AN IMPEDANCE METER (ZMeter)

Even though an impedance meter does not exist in practice, one can be simulated in PSpice for Windows. The purpose of this section is to design an impedance meter named **ZMeter**. ZMeter will be placed in the **Brum.slb** library. Section B.5 will illustrate how to use ZMeter.

First we will open the library editor and the **special.slb** library. Click **File/Edit Library /File/Open**. Scroll down until **special.slb** is visible. Click **special.slb/OK**. Next, we will copy and modify the symbol **VPRINT2**. Click **Part/Copy**, scroll down to **VPRINT2** and click on it. We will now change the name of the symbol we are about to make. Click inside the **New Part Name** box and delete the contents. Type *ZMeter*. Click **OK**. Next, we will display and move the pin numbers to more suitable locations. Double click the wire of the left pin and click **Display Name/OK**. Click the number 1 to select it and drag it to a position just to the left of pin 1 (pin on left). Double click the wire of the right pin and click **Display Name/OK**. Click the number 2 to select it and drag it to a position just to the right of pin 2 (pin on right). Selecting the number 2 may be somewhat difficult if the number is on the printer symbol. If you have difficulty selecting the number 2, move some of the lines around the number 2. After the number 2 is selected and moved, return the other lines to their original location. We will now make pin 2 the reference pin. Click **Graphics/Origin**. Place the pencil on the x of pin 2 and double click.

Next, we will change some attributes to make our impedance meter work properly. Click **Part/Attributes**. We will now change the name of the symbol. Click **REFDES**. Click inside the **Value** box and delete its contents. Type *ZMeter*. Click **Display Value/Save Attr**. We will now insert a one amp zero degree current source inside our symbol. This current source will force current out of pin 1. Click **TEMPLATE=**. Click inside the **Value** box at the very beginning in front of the first character. Type:*I^@REFDES<spacebar>%2<spacebar>%1<spacebar><spacebar>1 <spacebar>\n<spacebar>R^@REFDES<spacebar>%2<spacebar>%1 <spacebar>100meg<spacebar>\n* leaving spaces only where <spacebar> appears. The rest of the line stays as is. Click **Save Attr**. Next, we will make ZMeter display the magnitude, angle, real part and imaginary part of the impedance. Click **AC=**. Click inside the **Value** box and type **OK/Save Attr**. Repeat for **MAG=**, **PHASE=**, **REAL=** and **IMAG=**. When you are finished, there should be an OK after the equal signs on each of these lines. Click **OK**. We will now save the new part. Click **Part/Save Changes/Part/Save to Library**. Scroll down until **brum.slb** is visible. Click **brum.slb/OK**. Click **File/Return to Schematics**. If asked "Save changes to current library?" click **no**.

The impedance meter named ZMeter is now available in the **brum.slb** library. Pin 2 of ZMeter should be grounded for correct impedance reading.

## B. 5  USING ZMeter

The use of **ZMeter** is best illustrated by example. Figure B-4 shows the circuit whose total impedance was found in section 4.11 on page 61. In chapter 4 the total impedance was found by calculations as (2-j3.229)kΩ or 3.798kΩ at -58.228 degrees. Figure 4-7 shows the circuit set up using Brumgnachs one amp rule to find the total impedance. The total impedance using this method was (2-j3.226)kΩ or 3.797kΩ at -58.20 degrees. Here we are going to use our newly designed impedance meter **ZMeter** to find the same total impedance.

Use schematics to draw the circuit shown in Figure B-4 and insert the **ZMeter** as shown. Make sure pin 2 is grounded. To use **ZMeter** the frequency at which the impedance is desired must be specified. To set up the analysis for 1kHz click **Analysis/Set Up/ac Sweep/Linear**. In the **Sweep Parameters** box click inside the **Total Pts** box and change the contents to *1k*. Click inside the **Start Freq** box and change its contents to

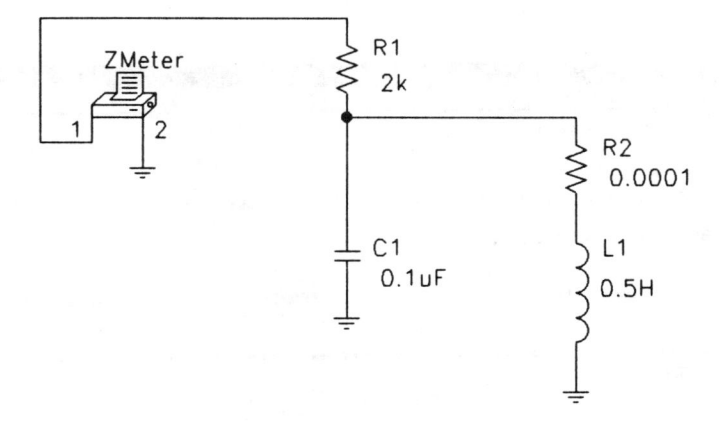

APPENDIX B   Making an Ohmmeter/Impedance Meter          127

**B-4   Impedance Measurement with ZMeter**

**1k**. Click inside the **End Freq** box and change its contents to *1k*. Click **OK**. Click inside the **enable** check box for the **ac Sweep** to enable the sweep (the box should have an X in it). Click **OK**. To prevent probe from running automatically, click **Analysis/Probe Set Up/Do Not Auto-run Probe**. To find the total impedance click **Analysis/Run PSpice**. When the analysis is finished, close the PSpice window. To see the results, click **Analysis/Examine Output** and scroll down until the ac analysis section is on the screen. The polar and rectangular forms of the total impedance are shown as VM (magnitude), VP (phase), VR (real part) and VI (imaginary part). From Figure B-5 these values are: 3.795kΩ at -58.2 degrees and (2-j3.226)kΩ. These values agree with the answers obtained in chapter 4.

## IB.6   MAKING ZMeter INTO ZProbe

The ZMeter symbol requires two connections when placed in a circuit. Pin 1 should be connected to the point in the circuit where the impedance value is desired and pin 2 should be connected to ground. ZMeter can be modified so that only one connection to the circuit to be measured is required. The ground connection can be included in the symbol so that the newly named **ZProbe** now only needs one connection to the circuit to be measured. We will now modify ZMeter and name it ZProbe.
   C l i c k   **File/Edit Library/File/Open/brum.slb/OK**.   C l i c k **Part/Copy/ZMeter**. Click inside the **New Part Name** box and delete its

APPENDIX B    Making an Ohmmeter/Impedance Meter

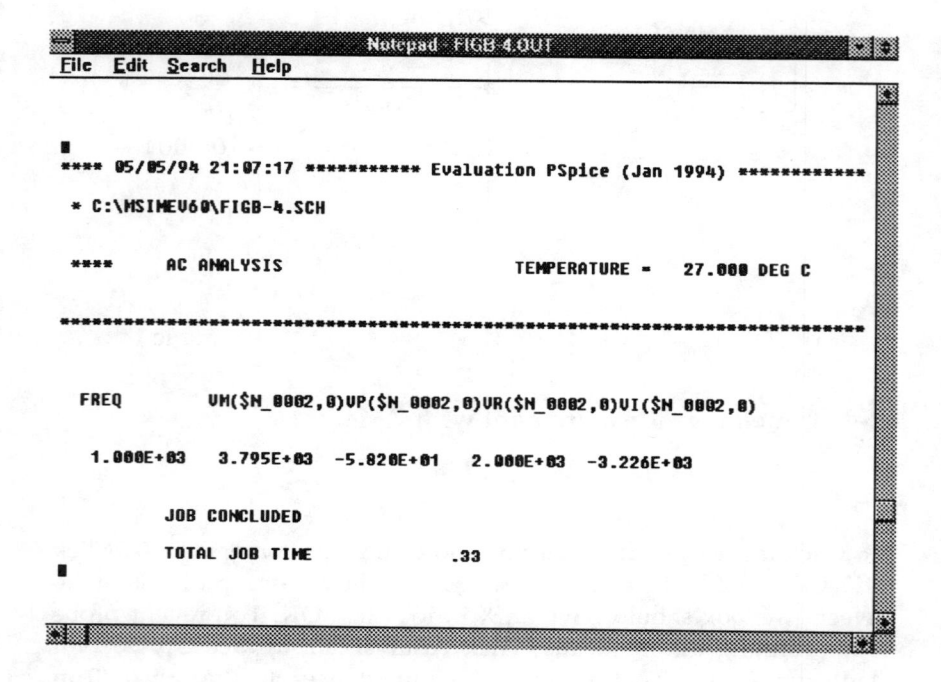

```
**** 05/05/94 21:07:17 ************ Evaluation PSpice (Jan 1994) ************
* C:\MSIMEV60\FIGB-4.SCH

****     AC ANALYSIS                      TEMPERATURE =   27.000 DEG C

********************************************************************************

  FREQ       VM($N_0002,0)VP($N_0002,0)VR($N_0002,0)VI($N_0002,0)

 1.000E+03   3.795E+03  -5.820E+01   2.000E+03  -3.226E+03

         JOB CONCLUDED
         TOTAL JOB TIME            .33
```

**B-5  Output of ZMeter**

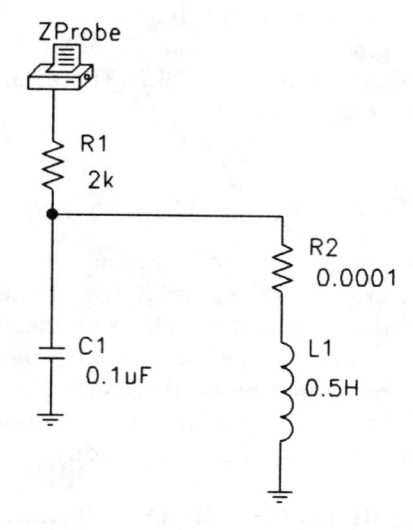

**B-6  Impedance Measurement with ZProbe**

## APPENDIX B    Making an Ohmmeter/Impedance Meter

contents. Type *ZProbe*. Click **OK**. Next we will connect pin 2 to ground and hide it. Double click pin 2. Click inside the **Hidden** check box to show an X. Click inside the **Net** box and type the number 0. Click **OK**. Click on pin 1 and drag it to the center bottom of the printer symbol. Double click pin 1 and click **Display Name/OK**. Click **Part/Attributes/REFDES**. Click inside the **Value** box and delete its contents. Type *ZProbe*. Click **Save Attr/OK**. Click **Part/Save Changes/Part/Save to Library/brum.slb/OK/File/Return to Schematic**. The one terminal impedance probe named ZProbe is now available from the brum.slb library. ZProbe needs only one connection in the schematic diagram and is used exactly like ZMeter. Figure B-6 shows ZProbe connected to a circuit.

# INDEX

Admittance, 59–60. See also $\overline{Y}$

Biasing, 91
BJT
  small signal amplifier, 92–93
  with voltage divider bias, 91–92
Brumgnach's 1 amp rule, 61, 123

Capacitor
  charging time constant, 84
  impedance of, 55–56
  impedance of
    with AC voltage source, 56
    with sinusoidal voltage source, 68–70
Circuit analysis, 38
  loop analysis, 39–41
  node analysis, 43–45
  Norton's equivalent circuit, 45–50
  superposition, 41–43
  Thevenin's equivalent circuit, 45–50
Circuit diagram
  analyzing circuit, 13
  changing data in name box, 13
  DC ammeter, 12
  DC voltage measurement, 12–13
  ground symbols, 12
  hard copy for, 13
  placing text on, 9
  saving, 13
  ten volt DC source selection, 11–12
  VSRC access, 11
  wires in, 12
Circuits
  digital, 99–104
  electronic, 87–98
  impedance with AC sources, 51–64
  resistive with DC sources, 29–50
  RLC
    with sinusoidal sources, 65–82
    with VPULSE, 83–86
Clicking, 5
Complex numbers
  and phasors, 52–53
Conductance, 32. See also G
Constant DC current sources, 20
Constant DC voltage sources, 20
Current, 19. See also I
Current divider rule, 34

Design Center for Windows
  drawing a circuit diagram, 10–13
  first exercise, 5–10
  menu selection and clicking, 5
  programs, 1–2
  Schematics Editor
    dialog boxes, 4
    menu bar, 3
    mouse, 4–5
    scroll bars, 3–4
    status bar, 4
    title bar, 2–3
  software installation, 1
  system requirements, 1
  VSRC symbol, modifying to look like a battery, 13–17
Dialog boxes, 4
Digital circuits

digital stimulus source (STIM1) and TTL inverter, 99–100
   modulo 3 synchronous counter, 102–4
   TTL 7408 and gate logic, 100
Display Options, 6
Double click, 5

Electronic circuits
   BJT small signal amplifier, 92–93
   BJT with voltage divider bias, 91–92
   full wave bridge rectifier, 88
      with capacitor filter, 88, 91
   half wave rectifier, 87–88
   JFET small signal amplifier, 93, 95
   tone control, 95, 98

Filter Design, 2
Final time, 83
Frequency domain, 53
Frequency response, 95, 98
Full-wave bridge rectifier, 88
   with capacitor filter, 88, 91

G, 32
Gate logic, 100–101
Get Part dialog box, 7

Half-wave rectifier, 87–88

I, 20
   finding, given V and R, 20–21
   finding corresponding values of
      given several values of R with V constant, 23–25
      given several values of V with R constant, 22–23
Impedance, 55-64. *See also* $\overline{Z}$
Impedance circuits
   of capacitor, 55–56
   of inductor, 57–58
   of resistor, 53–55
   in series, 58–61
   in series-parallel, 61–64

for Version 5.3, 105–19 (Appendix A)
INCLUDE
   capacitor with AC voltage using, 111
   impedances
      in parallel with AC voltage using, 114–16
      in series-parallel with AC voltage using, 116–19
      in series with AC voltage using, 113–14
   inductor with AC voltage using, 112–13
   resistor with AC voltage using, 107–9
Inductor
   impedance of, 57
      with AC voltage source, 57–58
      with sinusoidal current source, 70–72
$I_{peak}$, 69
IPRINT
   impedance of AC resistor using, 53–55
IR1, 34
$I_{R1}$, 60
IR2, 34, 37
IR3, 37
ISIN, 65
ISRC
   as phasor current source, 53
IT, 36
$I_T$, 60–61, 63

JFET small signal amplifier, 93, 95

Kirchhoff's Current Law, 34
Kirchhoff's Voltage Law, 31

Loop analysis, 39–41

Menu bar, 3
Menu selection, 5
Modulo 3 synchronous counter, 102–4

INDEX 133

Mouse, 4–5

Node, 32
Node analysis, 43–45
Norton's equivalent circuit, 45–50
NPN BJT amplifier, 92–93

Object selection, 5
Ohm's law, 19
Opening a file, 10–11

P, 20
Parallel resistive circuit
   with DC current source, 32–35
   with DC voltage source, 35
Parts, 2
Phase angle, 51–52
Phasor notation, 51–52
Phasors
   and complex numbers, 52–53
Pointing, 5
Polar form, 52
Power, 19–20. *See also* P
Print step, 83
PROBE, 2
   capacitor with sinusoidal voltage using, 68–70
   impedance of AC resistor using, 55, 109–10
   inductor with sinusoidal current using, 70–72
   series RC circuit with sinusoidal voltage using, 73–75
   series RLC circuit with sinusoidal voltage using, 79–82
   series RL circuit with sinusoidal voltage using, 75–79
Product over sum formula, 33, 60
PSpice, 2
Pulse voltage source
   RC circuit with, 83–85
   RLC circuit with, 86

RC circuit

   with VPULSE, 83–85
Reactance
   Capacitive, 55. *See also* $X_C$
   Inductive, 57. *See also* $X_C$
Rectangular form, 52
Resistance, 19. *See also* R
Resistors
   impedance of, 53–55
   in parallel
      with DC current source, 32–35
      with DC voltage source, 35
   in series
      with DC voltage source, 29–32
   in series-parallel
      with DC voltage source, 36–37
      with sinusoidal voltage source, 65–68
RLC circuits
   with pulse voltage source, 83–86
   with sinusoidal voltage sources, 65–82
RT, 29, 33
RTH, 46. *See also* Thevenin's equivalent circuit

S, 32, 59. *See also* Siemens
Saving a file, 10
Schematics, 1
   changing
      component's name, 9
      component's value, 9
   deleting components, 9–10
   drawing a circuit diagram, 11–13
   flipping a component, 8
   getting components from library, 7
      by typing their acronyms, 8
   moving position of component name or value, 9
   opening and opening a file, 10–11
   placing components, 7–8
   placing text on circuit diagram, 9
   rotating a component, 8
   saving files and closing, 10
   selecting and moving
      more than one component, 8

one component, 8
Schematics Editor, 2
Schematics Editor Window, 2–4
　display options, 6
Scroll bars, 3–4
Series-parallel circuit
　with DC voltage source, 36-
Series RC circuit
　with sinusoidal voltage source, 73–75
Series resistive circuits, 29–32
Series RLC circuit
　with sinusoidal voltage source, 79–82
Series RL circuit
　with sinusoidal voltage source, 75–79
Siemens, 32, 59
Sinusoidal current source, 65
Sinusoidal voltage sources
　capacitor with, 68–70
　inductor with, 70–72
　parameters of, 65
　resistor with, 65–68
　series RC circuit with, 73–75
　series RLC circuit with, 79–82
　series RL circuit with, 75–79
Sinusoidal waveforms, 51
Software installation, 1
Status bar, 4
STIM1, 99–100
Stimulus Editor, 2
Superposition, 41–43
System requirements, 1

Thevenin's equivalent circuit, 45–50
Timing diagram, 104
Title bar, 2
Tone control, 95, 98
Total admittance, 60
Total conductance, 33
Total impedance, 58
Total resistance, 33
Total response, 73, 75, 79
　forced component, 73, 75–76, 79–80

natural component, 73, 75–76, 79–80
Transient analysis setup, 83, 86
TTL 7404, 99
TTL 7408, 100–101

V, 20, 33
　finding, given I and R, 25–26
　finding corresponding values of
　　given several values of I with R constant, 26
　　given several values of R with I constant, 27–28
Voltage, 19. *See also* V
Voltage divider bias circuit, 91–92
Voltage divider rule, 31
Voltage drop, 30
Voltage rise, 31
$V_{peak}$, 71
VPRINT1, 53
VPULSE
　parameters of, 83
　RC circuit with, 83–85
　RLC circuit with, 86
VSIN, 65
VSRC
　accessing, 11
　modifying to look like a battery, 13–17
　as phasor voltage source, 53
VTH, 49. *See also* Thevenin's equivalent circuit

Watts, 19
Waveforms, 105–6

$X_C$, 55–56, 68. *See also* Capacitive Reactance
$X_L$, 57, 71. *See also* Inductive Reactance

$\overline{Y}$, 59. *See also* Admittance

$\overline{Z}$, 57–58. *See also* Impedance